Courses of Action for Enhancing U.S. Air Force "Irregular Warfare" Capabilities

A Functional Solutions Analysis

Richard Mesic, David E. Thaler,
David Ochmanek, Leon Goodson

Prepared for the United States Air Force
Approved for public release; distribution unlimited

RAND PROJECT AIR FORCE

The research described in this report was sponsored by the United States Air Force under Contract FA7014-06-C-0001. Further information may be obtained from the Strategic Planning Division, Directorate of Plans, Hq USAF.

Library of Congress Cataloging-in-Publication Data

Courses of action for enhancing U.S. Air Force "irregular warfare" capabilities : a functional solutions analysis / Richard Mesic ... [et al.].
p. cm.
Includes bibliographical references.
ISBN 978-0-8330-4874-5 (pbk. : alk. paper)
1. United States. Air Force—Operational readiness. 2. Asymmetric warfare—United States. 3. Unified operations (Military science)—United States.
I. Mesic, Richard, 1943–

UG633.C688 2010
358.4'142—dc22

2010000854

Cover photo courtesy of U.S. Air Force/Senior Airman Randi Flaugh.

Published 2010 by the RAND Corporation
1776 Main Street, P.O. Box 2138, Santa Monica, CA 90407-2138
1200 South Hayes Street, Arlington, VA 22202-5050
4570 Fifth Avenue, Suite 600, Pittsburgh, PA 15213-2665
RAND URL: http://www.rand.org/
To order RAND documents or to obtain additional information, contact
Distribution Services: Telephone: (310) 451-7002;
Fax: (310) 451-6915; Email: order@rand.org

Preface

This monograph is the product of a "quick-turn" study to assist U.S. Air Force leadership in choosing ways to enhance Air Force capabilities and capacities for irregular warfare (IW). The primary sponsor was the Vice Commander of Air Combat Command. The Director of the Air Force Quadrennial Defense Review (Office of the Assistant Vice Chief of Staff) and the Director of Operational Capability Requirements (Office of the Deputy Chief of Staff for Operations, Plans and Requirements) had corollary responsibility.

The work documented here draws from a large body of recent RAND research, as well as lessons-learned studies and other assessments done within the Department of Defense, and also from intensive discussions with Air Force subject-matter experts. Although there have been numerous discussions along the way (and in past work) about alternatives and their relative merits, this monograph focuses on conclusions. In some cases, details should be further checked, but the basic elements of our conclusions appear to us sound. The research was conducted within the Strategy and Doctrine Program of RAND Project AIR FORCE in collaboration with Dr. Leon Goodson of STR, L.L.C.

RAND Project AIR FORCE

RAND Project AIR FORCE (PAF), a division of the RAND Corporation, is the U.S. Air Force's federally funded research and development center for studies and analyses. PAF provides the Air Force with independent analyses of policy alternatives affecting the development, employment, combat readiness, and support of current and future aerospace forces. Research is conducted in four programs: Force Modernization and Employment; Manpower, Personnel, and Training; Resource Management; and Strategy and Doctrine.

Additional information about PAF is available on our Web site: http://www.rand.org/paf

STR, L.L.C.

STR, L.L.C. is a Virginia-based company that provides professional services and management consulting to customers in both the private and government sectors.

Additional information about STR, L.L.C. is available on its Web site: http://www.strllc.com

Contents

Figures

Tables

Summary

This monograph documents a "quick-turn" study to assist Air Force leadership in determining actions to enhance USAF capabilities and capacities for joint and combined operations in irregular warfare. The Secretary of Defense has directed the military services to come forward with operational and force structure initiatives to rebalance their forces across IW and conventional-warfare mission areas. According to this direction, U.S. forces must be prepared to conduct IW operations for many years to come. Airpower plays critical roles in IW operations that are both direct—e.g., hunting down and attacking terrorist groups—and indirect—e.g., training, equipping, advising, and assisting partners.

As the Air Force enhances its IW capabilities it must also maintain capabilities for deterring or defeating regional or near-peer adversaries in large-scale conflicts. In rebalancing, it must recognize that, although IW and large-scale combat operations have features in common, IW operations are not a lesser included case of major combat operations. This is particularly apparent when one recognizes that IW is largely about influencing relevant populations and operating by, with, and through partner nations. Further, military force is only one of the relevant instruments of national power and may even be secondary to political and economic instruments in many contexts.

Courses of Action for Enhancing Air Force IW Capabilities

We identify four dimensions in which the Air Force could act to expand its contributions to joint IW operations, both direct and indirect. These dimensions involve

- adapting the institutional Air Force to embrace an IW mindset even as it maintains capabilities for traditional missions
- enhancing its capacity to meet the pressing needs of commanders in Iraq and Afghanistan
- more generally, ensuring that it can meet *global* demands to support partner nations against extremist threats to their stability and populations' well-being
- building force-structure capabilities and capacities to sustain this effort over the years of the "long war."

The courses of action (CoAs) we propose address each of these dimensions. We formulate our CoAs using an incremental approach. CoA 0 is the bedrock, and should be taken in any case. The additional CoAs are thematically coherent incremental building blocks. Each has merit and each would build on the others. All are grounded in analysis, recent operational experience of USAF general-purpose forces (GPF) in IW, and long experience by USAF special operations forces (SOF). Although there have been many discussions along the way (and in past work) about alternatives and their relative merits, this monograph focuses on conclusions. In some cases the initiatives have not been subjected to detailed force-effectiveness analysis. In many cases, the Air Force itself is the source of initiatives. In fact, most of the initiatives identified here are already being considered and, in some instances, pursued by USAF organizations. Nevertheless, while in all cases the details should be further checked and refined, the basic elements appear to us sound.

Course of Action 0: Set the Climate

Serving as the foundation for all other courses of action, CoA 0 focuses on deepening the institutional commitment to developing IW expertise in the USAF and providing relevant capabilities. In our view, CoA 0 is not optional, but essential, if the Air Force undertakes to truly embrace irregular warfare as a USAF competency. It is for this reason that we call it "CoA 0." It includes five initiatives to inculcate an IW mindset in USAF culture and to ensure that IW has a top priority:

- early and frequent leadership emphasis on IW as an institutional priority
- a permanent, high-level management organization to monitor and direct IW activities in the Air Force with unity of purpose
- realignment of career management to encourage and reward IW experience
- realignment of education and training to greatly increase the exposure of airmen to IW concepts throughout their careers
- expansion of the Coalition Irregular Warfare Center (CIWC) as a USAF focal point for IW-related concept development, technology exploitation, and strategic assessment.

We estimate the resources to implement the initiatives of CoA 0 as 180 personnel, an initial investment of about $7 million (mostly for office equipment), and an annual operations and support (O&S) cost of about $30 million. (See pp. 17–20.)

Course of Action 1: Succeed in Iraq and Afghanistan

Course of Action 1 is predicated on the judgment that the Air Force should prepare for continued heavy involvement in direct and indirect operations in Iraq and Afghanistan. Activities in Afghanistan have recently increased and are likely to continue for years. The rugged terrain and lack of roads, and the distances between distributed battle areas, will highlight the flexibility and reach of airpower. In addition, recent decisions to greatly expand the size of the Afghan National Army presage a long-term requirement for the USAF to help build the Afghan Air Corps and support the extension of the Afghan government's reach into rural areas. The situation in Iraq is more ambiguous. The U.S.-Iraq security agreement calls for U.S. forces to withdraw from Iraq by 2011. That may well occur on schedule, but it is also possible that Iraqi leadership will ask U.S. forces to remain with a smaller, less-visible presence focused on building Iraqi security forces while providing niche capabilities in support of Iraqi operations. These niche capabilities could well include USAF assets for close-air support (CAS), armed overwatch, intelligence, surveillance, and reconnaissance (ISR), airlift, command and control, force protection, and other important functions—perhaps for 5–10 years.[1]

Thus, CoA 1 addresses shortfalls in current and potential operations in Iraq and Afghanistan by pushing more Air Force capability and capacity forward, preparing the USAF for what could be a longer-than-hoped-for stay while broadening efforts dedicated to establishing the Iraqi Air Force and Afghan Air Corps as strong, self-sustaining entities. The intent of CoA 1 is to show immediate effects in the Iraqi and Afghan theaters and to field systems that improve the effectiveness and efficiency of deployed forces for operations.

The initiatives for CoA 1 involve

- training more Joint Terminal Attack Controllers (JTACs) and embedding specialized air expertise at multiple echelons to support planning and operations
- providing additional ISR and transferable CAS and mobility capability
- pushing forward a range of capabilities for information operations, strategic communications, agile combat support, and lessons-learned capacity
- providing a more robust, tailorable Air Education and Training Command (AETC) pipeline for training USAF advisors
- adding an Air Force Special Operations Command (AFSOC) squadron of combat aviation advisors

---

[1] The commander of Multi-National Forces in Iraq, General Ray Odierno, has stated his belief that U.S. air capabilities could be needed in Iraq for another five to ten years. More recently, Iraqi Prime Minister Nuri al-Maliki left the door open to continuing U.S. military presence after 2011: "If Iraqi forces need more training and support, we will reexamine the [status of forces] agreement at that time, based on our own national needs." See Department of Defense, "DoD News Briefing with Lt. Gen. Odierno from Iraq," January 17, 2008; and "A Conversation with Iraq's Prime Minister Nuri al-Maliki," United States Institute of Peace, July 23, 2009.

- creating a regional air academy focused on Iraqi and Afghan airmen.

Implementing all the CoA 1 initiatives would require an estimated 3,600 personnel, 200 aircraft (including additional intelligence-gathering MC-12s, the conceptual OA-X light attack platform, and a family of light cargo aircraft), about $1.9 billion in initial investment (mainly for aircraft), and some $423 million in yearly O&S costs. (See pp. 20–27.)

Course of Action 2: Support Partners Globally

Notwithstanding the importance of operations in Iraq and Afghanistan, they are by no means archetypal scenarios for future IW. The challenges are global, and operations in other contingencies are unlikely to look much like those in Iraq and Afghanistan. CoA 2 would recognize this global challenge. It would enable the United States and its partners to put unremitting pressure on terrorists and insurgents around the world—working largely by, with, and through partners with diverse needs and at varied developmental levels—while enhancing its contribution to discreet U.S. direct operations against terrorist groups. This global demand for the expertise of U.S. airmen is neither static nor well defined as of yet, but its magnitude is widely recognized as being well beyond current U.S. capacity.

As the USAF presses ahead with its efforts to build Iraqi and Afghan air capabilities, CoA 2 addresses these global demands through initiatives that

- establish IW advisory wings in the general-purpose force and in AFSOC
- expand AETC training efforts to sustain the larger advisory force
- pursue advisory and employment concepts for agile combat support
- embed air advisory elements in air components of the combatant commands (COCOMs)
- establish additional regional air academies

- deploy transferable, multi-mission (including ISR) light utility aircraft to the COCOMs
- expand USAF human-intelligence capabilities.

Overall, CoA 2 involves 3,000 personnel, 255 aircraft, an investment of $2.3 billion, and an annual O&S cost of $374 million. (See pp. 27–31.)

Course of Action 3: Ensure Future Access and Action

Course of Action 3 is aimed at long-term force structure rebalancing to support IW. Its main objectives are to ensure that the Air Force would be positioned to support surge IW operations in the future and that special operations forces could operate in less permissive environments, at times clandestinely. We divide CoA 3 into two elements, each associated with one of these objectives. In the first element, units equipped with an additional 93 manned ISR platforms (to complement what we assume will be continued expansion of unmanned systems) and 300 counterinsurgency-dedicated CAS platforms would be available for global operations. We estimate the resources needed to pursue these two initiatives at 4,400 personnel, $4.7 billion in investment, and $600 million in annual O&S costs (for all 393 aircraft).

The second CoA 3 element enhances special operations capabilities in the out-years with a low-observable, next-generation gunship concept and a low-observable SOF mobility platform for infiltration/exfiltration operations in denied areas and over long distances. This element of CoA 3 recognizes that even as the United States emphasizes operating by, with, and through partners, there will still be a need for U.S. forces to conduct direct operations against nonstate adversaries. We estimate resources for the 48 aircraft associated with these initiatives to be 1,320 personnel, $20.5 billion in investment, and $216 million in annual O&S costs. Most of the resources would be needed beyond 2020, although some research, development, test, and evaluation (RDT&E) would appear late in the Future Years Defense Plan. (See pp. 31–33.)

The matrix in Table S.1 provides a summary of our proposed CoAs and initiatives, grouped by functional area—institutional, operational capability, building partner capacity, and future procurement. An initiative identified as part of a CoA receives a checkmark under that CoA. However, initiatives may also play roles in follow-on CoAs. As such, if an initiative also supports the goals or emphases of other CoAs, it receives a plus sign under those CoAs. Estimated resources for each CoA appear at the bottom of the matrix. For CoA 3, resources are divided between the first two (MC-12 and OA-X) and second two initiatives (gunship and mobility platform, which are bracketed). The total incremental manpower, total aircraft inventory, investment costs, and annual O&S costs refer to fully implemented initiatives. Appendix B details the breakout and flow of these manpower and cost figures.

Table S.1
Summary of Initiatives to Support Courses of Action

Initiatives	CoA 0 Set the Climate	CoA 1 Succeed in Iraq and Afghanistan	CoA 2 Support Partners Globally	CoA 3 Ensure Future Access and Action
Institutional				
Provide leadership emphasis	√	+	+	+
Designate HAF integrator for IW	√	+	+	+
Emphasize IW career incentives	√	+	+	+
Expand IW curricula in education and training	√	+	+	+
Expand CIWC	√	+	+	+
Operational Capability				
Embed air expertise forward		√		
Add JTACs		√		
Push and sustain airborne ISR and PED		√	+	√
Emphasize innovative information operations		√	+	+
Synchronize strategic communications		√		
Mitigate agile combat support shortfalls		√	√	+
Operationalize analyses		√	+	+

Table S.1 (continued)

Initiatives	CoA 0 Set the Climate	CoA 1 Succeed in Iraq and Afghanistan	CoA 2 Support Partners Globally	CoA 3 Ensure Future Access and Action
Building Partner Capacity				
Expand advisor training		√	√	+
Stand up GPF advisor unit and elements			√	+
Add combat aviation advisors		√	√	+
Expand USAF human intelligence			√	+
Build regional air academies		√	√	+
Future Procurement				
Provide transferable, COIN-dedicated CAS and armed overwatch platform		√	+	√
Develop and procure light cargo aircraft		√	√	+
Develop and deploy next-generation gunship				√
Develop and deploy next-generation SOF mobility platform				√
Total Estimated Resources				
Manpower	181	3,594	2,995	4,395 [1,320]
Total aircraft inventory	0	200	255	393 [48]
Investment	$7 million	$1.9 billion	$2.3 billion	$4.7 billion [$20.5 billion]
Annual O&S	$30 million	$423 million	$374 million	$598 million [$216 million]

Concept Development for Irregular Warfare

Although this monograph has focused on macro IW issues, such as leadership, culture, and human capital, the Air Force is also a leader and innovator in applying high technology to challenging, complex problems. Emerging operational concepts needed to utilize such technology have much to offer in IW operations. The USAF needs to be at the forefront of developing concepts for ISR (especially signals intelligence and persistent wide-area surveillance), space and cyberspace capabilities, communications, and low-collateral-damage weapons. Moreover, because building partner capacity and operating indirectly are so central to IW, the Air Force should pursue a dedicated effort to make the fruits of these technologically driven concepts available to partners within export control parameters and without compromising sensitive or classified U.S. capabilities. In fact, concept development activities should explicitly consider applicability to capacity-building and foreign internal defense. It would be the responsibility of the IW integrator at Headquarters Air Force and the expanded Coalition Irregular Warfare Center to ensure that technological opportunities for novel IW applications were explored, developed, and fielded. (See pp. 38–40.)

Conclusions

In summary, the Secretary of Defense has directed the Air Force and the other services to adapt to the emerging threat environment characterized as the "long war." The direction is clear, but the specifics present numerous challenges, not the least of which is the need to begin to change the USAF culture from one focused on the challenges of major combat operations (challenges that are not going away) to one equally accomplished in irregular warfare. The Air Force has a great deal to offer in this realm—indeed, U.S. efforts to counter terrorism and insurgency, support partners in foreign internal defense, and bring stability to vulnerable populations *could not be done without the Air Force.* In this spirit, we formulate and propose the CoAs in this report.

There is a compelling rationale for immediately beginning to implement the initiatives in CoAs 0, 1, and 2—or some versions thereof.[2] The need to adapt as an institution, to redouble efforts in Afghanistan and to support an orderly withdrawal and potential post-withdrawal presence in Iraq, and to pressure adversaries and support partners around the world is unassailable. As the Air Force implements these CoAs and gains additional experience in, and knowledge about, IW operations, USAF leadership can take more time to consider the more ambitious and costly initiatives proposed in CoA 3. (See pp. 41–43.)

Finally, while there are material initiatives in these CoAs, most of the initiatives are nonmaterial. People, not platforms, are the key to transforming the Air Force and rebalancing its IW and conventional capabilities, allowing it to work more effectively and efficiently by, with, and through partner nations in the long war. As Secretary of Defense Robert Gates wrote, "In the end, the military capabilities needed cannot be separated from the cultural traits and the reward structure of the institutions the United States has: the signals sent by what gets funded, who gets promoted, what is taught in the academies and staff colleges, and how personnel are trained."[3]

---

[2] The analysis herein provides rough estimates of resources required to implement initiatives, but it does not attempt to identify where those resources might originate (i.e., either from offsets within the Air Force program or from external funding sources).

[3] Robert M. Gates, "A Balanced Strategy: Reprogramming the Pentagon for a New Age," *Foreign Affairs*, January/February 2009.

Acknowledgments

The authors wish to thank a number of people for their support of the study documented in this monograph.[4] Maj Gen R. Mike Worden, Vice Commander of Air Combat Command (ACC/CV), and Maj Gen Frank Gorenc, ACC Director of Air and Space Operations (ACC/A3), sponsored the study; Lt Col J. David Torres-Laboy was our ACC point of contact and arranged the Air Combat Command IW Workshop in October where we consulted with a broad range of USAF experts. Maj Gen James Hunt, Director, Air Force Quadrennial Defense Review, Office of the Assistant Vice Chief of Staff, Headquarters Air Force (AF/CVAQ), and Maj Gen Stephen Mueller, Director, Operational Capability Requirements, Deputy Chief of Staff for Operations, Plans and Requirements, Headquarters Air Force (AF/A5R), had corollary responsibility for the study, provided insights, and sponsored the briefing through Headquarters Air Force. Special thanks go to Carl Rehberg, Chief, Analysis Division (AF/CVAQ), and George Monroe (AF/CVAQ, SAIC) for the hours they spent with us focusing the study and honing the report, and to Col Charles Sachs, Deputy Director, IW Requirements (AF/A5R-Q), and Lt Col Justin Bobb (AF/A5R-Q) for helping us synchronize our study with parallel efforts of the IW Task Force chaired by Maj Gen Mueller. We would also like to express our appreciation to Andrew Hoehn, Director of Project AIR FORCE, for his encouragement of our work and the guidance he provided throughout the study.

[4] All ranks and positions were current at the time the research was performed, from October 2008 through January 2009.

Numerous other Air Force personnel dedicated time to share with us their expertise, insights, and operational experience. We are indebted to Lt Gen Daniel Darnell, Deputy Chief of Staff for Air, Space and Information Operations, Plans and Requirements, Headquarters Air Force; Maj Gen Kurt Cichowski, Vice Commander, Air Force Special Operations Command (AFSOC); Col Patricia Battles (AF/A4LX); Col Gary Crowder (AF/A3O-AY); Col Albert Elton (AFSOC/A8X); Col James Iken (AF/A7CX); Col Philip Senna (AETC/A3A); Lt Col Jonathan Dagle (NGB/A8X); Lt Col Peter Davies (AFWC/CIWC); Lt Col James Mueller (SAF/IARM); Lt Col Michael Pietrucha (ACC/A3F); Lt Col Craig Werenskjold (AFSOC/A91); Lt Col Martin Winkler (AF/A9AF); Maj Timothy Childress (AF/A5XS); Maj Matthew Rosen (AMC/A8XP); Maj Christopher Will (NGB/A3OC); Aryea Gottlieb (AFSOC/A8XC); G. Hale Laughlin (AFSOC/6SOS); Wesley Long (USAFCENT/A3T); Robyn Read (AETC AFRI/RIR); and Louis Finch (AF/A8XP, STR). We spoke with many others as well during the study, and we are grateful to all of them for the thoughts and ideas they shared. We also appreciate the review and comments provided by Col Curtis "Gator" Neal (USAF, Ret.) on our final draft.

Many colleagues at RAND helped us formulate, develop, and assess initiatives as well as frame the structure of the briefing. Many thanks go to John Ausink, Tony Bauernfeind, James Chow, Natalie Crawford, Paul Davis, Adam Grissom, Thomas Hamilton, Ryan Henry, Stephen Hosmer, Alexander Hou, Myron Hura, David Johnson, Kristin Lynch, Patrick Mills, Jennifer Moroney, Karl Mueller, Edward O'Connell, David Orletsky, Donald Stevens, and Alan Vick.

We also appreciate the considerable support we received from other members of the RAND staff. Communications analyst Jerry Sollinger refined versions of the briefing upon which this report is based. Project associate Aidan Winn drafted notes of discussions during a two-day meeting at RAND to kick off this effort. Administrative assistant Francisco Walter helped prepare the draft documented briefing that served as an earlier version of this monograph, and Jane Siegel prepared the final manuscript. Finally, Nikki Shacklett and Kimbria McCarty edited the monograph.

Many thanks also go to Lynn Davis for reviewing the final draft of this monograph, and to communications analyst Michael Neumann for transforming the documented briefing into a formal report. Finally, we thank our RAND peer reviewers, Heather Felton and Paul Davis, and our external peer reviewer, John A. Nagl of the Center for a New American Security. Their efforts greatly improved this document.

Responsibility for the content of this monograph lies solely with the authors.

Abbreviations

ACC	Air Combat Command
ACS	agile combat support
AE	air effects
AETC	Air Education and Training Command
AFSC	Air Force Specialty Code
AFSOC	Air Force Special Operations Command
AP	air planning
BPC	building partner capacity
CAA	combat aviation advisor
CAOC	combined air operations center
CAS	close air support
CIWC	Coalition Irregular Warfare Center
CoA	course of action
COCOM	combatant command
COIN	counterinsurgency
DoD	Department of Defense
GPF	general-purpose forces
HAF	Headquarters Air Force
HUMINT	human intelligence
I/O	information operations
ISR	intelligence, surveillance, and reconnaissance
IW	irregular warfare

JTAC	Joint Terminal Attack Controller
MILCON	military construction
NTISR	nontraditional ISR
O&M	operations and maintenance
O&S	operations and support
PAF	Project AIR FORCE
PED	processing, exploitation, and dissemination
RDT&E	research, development, test, and evaluation
SEIs	special experience identifiers
SOF	special operations forces
TACP	tactical air control party
TRADOC	[U.S. Army] Training and Doctrine Command
TTPs	tactics, techniques, and procedures
USAF	United States Air Force
USCENTCOM	U.S. Central Command
USJFCOM	U.S. Joint Forces Command
USSOCOM	U.S. Special Operations Command

CHAPTER ONE
Introduction

Terrorist and insurgent groups motivated by extremist ideologies pose a serious threat to U.S. interests and must be actively countered. The United States must conduct efforts to defeat such groups over the long term and apply multiple dimensions of its national power. Such efforts will include surveillance and intelligence assessment, sustained foreign internal defense, and direct counterterrorism and counterinsurgency operations—activities that are collectively referred to as "irregular warfare" (IW). "Waging protracted irregular warfare depends on building global capability and capacity" and thereby requires regional and cultural acumen to affect indigenous populations and governments. The struggle against extremist groups "will not be won by the United States alone."[1]

Despite consensus about the nature of the problem and the strategy for meeting it, neither the Department of Defense (DoD) nor other agencies of the U.S. government have yet structured themselves appro-

[1] Department of Defense, *Irregular Warfare (IW) Joint Operating Concept (JOC)*, Version 1.0, Washington, D.C., September 11, 2007, p. 1. According to the IW JOC, irregular warfare is defined as "a violent struggle among state and non-state actors for legitimacy and influence over the relevant populations. IW favors indirect and asymmetric approaches, though it may employ the full range of military and other capabilities, in order to erode an adversary's power, influence, and will" (p. 6). It encompasses "insurgency; counterinsurgency; unconventional warfare; terrorism; counterterrorism; foreign internal defense; stabilization, security, transition, and reconstruction operations; strategic communications; psychological operations;, information operations; civil-military operations; intelligence and counterintelligence activities; transnational criminal activities, including narco-trafficking, illicit arms dealing, and illegal financial transactions, that support or sustain IW; and law enforcement activities focused on countering irregular adversaries" (pp. 9–10).

priately for long-term effort. This is partly due to the pressing demands of the ongoing wars in Iraq and Afghanistan, which have made it difficult to focus on measures called for elsewhere in the world on a long-term basis. The recent success of stabilization efforts in Iraq, the impending drawdown of U.S. ground forces there, increasing demands in Afghanistan and elsewhere in the world, and DoD's Quadrennial Defense Review all provide guidelines for considering how joint forces should be organized, trained, equipped, and postured for future operations against terrorist and insurgent groups abroad.

For more than eight years after the attacks of September 11, 2001, U.S. policymakers have made clear that ensuring U.S. security will demand enduring efforts by DoD and other agencies to maintain pressure on the global extremist network. Hence, treating IW as an anomalous phenomenon that can be adequately dealt with through a series of ad hoc arrangements is no longer appropriate (if it ever was). Rather, it is time to structure portions of the force for these missions and to plan to employ them accordingly for many years to come.

Objectives and Approach

In response to direction from the Secretary of Defense, the United States Air Force (USAF) leadership plans in the near term to identify and implement a range of initiatives for enhancing the service's contributions to irregular warfare operations and to meet DoD guidance that directs its components to "recognize that IW is as strategically important as traditional warfare."[2] The Air Force asked RAND's Project AIR FORCE (PAF) to conduct a "quick-turn" study that would provide the leadership with a menu of actions it could consider both in the very near term and over an extended period to strengthen and expand the Air Force's capabilities to take part in joint and interagency efforts in irregular warfare.

---

[2] Department of Defense, *Irregular Warfare (IW)*, Washington, D.C.: Office of the Under Secretary of Defense for Policy, DoDD 3000.07, December 1, 2008.

This study is intended to support the Air Force's efforts to enhance its contribution to the U.S. capacity and capability to conduct irregular warfare. Rather than developing detailed endpoints for Air Force capabilities or structure in the future—e.g., what the USAF could look like in, say, 2015—we propose *solution vectors* that could enable the Air Force to move out quickly while continuing to consider responses to emerging demands in a dynamic IW environment.[3] Our goal is thus less ambitious and more circumspect. The environment in which current IW operations are taking place is very dynamic. Although relevant issues are being actively debated in a number of forums, it is clearly essential that the USAF better position itself to meet existing and emerging IW demands. In that spirit, we have attempted to develop robust vectors, rather than endpoints, for incremental solutions that address these demands. Every initiative proposed in this monograph, therefore, shares the following characteristics:

- It would provide, enhance, or expand a capability that is relevant to an enduring need of U.S. commanders charged with conducting IW operations.
- It is within the realm of the USAF's responsibilities under Title 10 of the U.S. Code.
- At the level of effort recommended here, there is little risk of the Air Force fielding *more* of the capability than will be called for. Rather, the opposite is likely to be true. But as the USAF gains experience with the newly enhanced capability, adjustments can be made to best meet a dynamic demand.

In developing the initiatives, PAF focused on the demands that new or extended Air Force capabilities could address. With a few excep-

---

[3] In this monograph we refer to "irregular warfare," the term used in the study's tasking, but we recognize that such terminology is controversial and may change. Some authors refer to "hybrid warfare" because operations in a theater may simultaneously include a mix of so-called "traditional" and "irregular" methods. More generally, some of the operations in question (e.g., surveillance and armed overwatch with occasional strikes) apply not only to "warfare" but also to post-major-combat circumstances during which the challenges are sometimes referred to as stability, security, transition, and reconstruction (SSTR).

tions, PAF was asked to forgo deriving organizational structures during the research.[4] Additionally, the analysis herein provides rough estimates of resources required to implement initiatives, but it does not attempt to identify where those resources might originate (i.e., either from offsets within the Air Force program or from external funding sources). For this study, PAF developed courses of action (CoAs) and associated initiatives in collaboration with STR, L.L.C.,[5] which estimated the potential costs of the initiatives over the Future Years Defense Plan and beyond. Because of the compact timelines for this study (approximately 60 days), we structured it fundamentally to gather, evaluate, refine, and compare work that has previously been done on IW operations and airpower's roles in them.

This study was initiated as an IW "functional solutions analysis" based on an existing body of functional area analysis/functional needs analysis in the form of an earlier PAF draft report.[6] Although the USAF adopted the draft report as the study's analytical point of departure, the study team recognized that to ensure the consideration of an appropriately broad mix of candidate initiatives, the effort must incorporate information and analyses from a wide range of other sources and subject matter experts, both in the Air Force and beyond, who had broad operational and institutional experience in irregular warfare.[7]

---

[4] One might imagine the development of an "IW AF," either as an extension of U.S. Air Force Special Operations Command (AFSOC) or as a new organizational entity dedicated to IW. We have not been able to address the pros and cons of major organizational shifts such as these. Instead, our approach here has been to address IW as a "total force" issue, leaving the macro organizational issue to future study once the IW challenges and opportunities (as outlined in our CoAs, for example) are better understood.

[5] STR, L.L.C. is a Virginia-based company that provides professional services and management consulting to customers in both the private and government sectors. Additional information about STR, L.L.C. is available on its Web site.

[6] See John Stillion et al., "Air Force Phase Zero Capabilities-Based Analysis: Functional Area Analysis and Functional Needs Analysis," unpublished RAND research, September 2007.

[7] The study team consulted experts at Headquarters Air Force, Air Combat Command, Air Mobility Command, Air Force Central Command, U.S. Air Force Special Operations Command, Air Education and Training Command, Air Force Reserve Command, and the Air National Guard.

This report provides the Air Force with a coherent direction. All of the proposed initiatives have merit because they are based on lessons learned over eight years of operational experience by USAF general-purpose forces (GPF) in irregular warfare and much longer experience by USAF special operations forces (SOF). However, the initiatives have not been subjected to detailed force-effectiveness analysis. This report does not provide the basis for determining the best allocation of resources across the USAF, or where the next marginal dollar should be spent for the best return on investment.

Organization of This Monograph

Looking ahead to the remaining chapters in this monograph, Chapter Two provides an understanding of how the United States views irregular warfare and the greater emphasis DoD has placed on IW in recent years. Chapter Three examines each of the four proposed courses of action in detail, and identifies each of the initiatives that correspond to each course of action. Chapter Four offers recommendations to help Air Force leaders successfully implement the various IW initiatives. Chapter Five wraps up the main body of the monograph by presenting our conclusions. Appendix A provides a detailed examination of each initiative, including the objectives, estimated costs, implementation timeline, and related employment and fielding concepts. Finally, Appendix B provides a more detailed look at the estimated resources necessary to implement the IW initiatives over time.

CHAPTER TWO

Engaging in Irregular Warfare

> The United States is a nation engaged in what will be a long war. Since the attacks of September 11, 2001, our Nation has fought a global war against violent extremists who use terrorism as their weapon of choice, and who seek to destroy our way of life. Our enemies seek weapons of mass destruction and, if they are successful, will likely attempt to use them in their conflict with free people everywhere. Currently, the struggle is centered in Iraq and Afghanistan, but we will need to be prepared and arranged to successfully defend our Nation and its interests around the world for years to come.[1]

So began the DoD *Quadrennial Defense Review Report* of February 2006. Similar statements appear in the opening pages of the 2006 *National Security Strategy* and the guidance and planning documents that support that strategy.

The United States is at war. It is a rallying cry for harnessing America's vast resources and focusing all its elements of national power—political, diplomatic, economic, ideological, and military—to prosecute a campaign that will bring "victory." And by terming this a "long war," the image emerges of a conflict that is expected to last years, even decades, due to the nature of the adversary. More recently, Secretary of Defense Robert Gates has emphasized that irregular warfare-related operations "cannot be considered exotic distractions or tem-

---

[1] Donald Rumsfeld, U.S. Secretary of Defense, *Quadrennial Defense Review Report*, Washington, D.C., February 6, 2006, p. v.

porary diversions."[2] He has clearly directed the military services to more aggressively address operational and force structure initiatives to respond to existing and emerging IW security challenges.

Thus, the U.S. armed forces must prepare to conduct these operations for many years to come. At any given time, these U.S. IW operations abroad will comprise a mix of "direct" and "indirect" efforts. In direct operations, U.S. forces will operate overtly or covertly, hunting down and attacking or disrupting terrorist and insurgent groups. These operations range from surveillance, to direct action, to routine patrols, such as the maritime intercept operations the U.S. Navy has been conducting off the Horn of Africa. Indirect operations, by contrast, are aimed at improving the effectiveness of host country forces. They center on efforts to train, equip, advise, and assist others, usually through intensive, hands-on training missions in-country to help host governments build capacity and legitimacy. In most cases U.S. forces will not take part in combat operations in the host country, though they may provide important supporting functions such as intelligence, surveillance, and reconnaissance (ISR), planning, tactical airlift, communications, and logistics. The defense strategy calls for a shift in the weight of effort over time from direct to indirect operations, reducing the U.S. footprint in the countries involved (which tends to delegitimize local governments) and enabling partners to maintain their own security and provide governance of their populations.[3] Airpower plays critical roles in direct and indirect IW operations: the Air Force should excel in both.

Conducting Air Operations in Irregular Warfare

Air Force assets are currently involved in a number of IW operations, including direct and indirect operations in Iraq, Afghanistan, the Phil-

---

[2] Robert Gates, U.S. Secretary of Defense, transcript of speech at the National Defense University, September 29, 2008.

[3] See Department of Defense, *National Defense Strategy*, Washington, D.C., June 2008, pp. 8–9.

ippines, and a number of other countries. One notable phenomenon in IW operations thus far—particularly direct operations in Iraq and Afghanistan—is that the effectiveness of Air Force operations in support of combatant commanders is often achieved by personnel who find better ways to do things, ways that get around burdensome and unproductive processes and constraints, with a focus on providing what the commanders need. As a result, many successful activities are "nondoctrinal." Typically, this means more decentralized decisionmaking, fewer constraints, crossing of stovepipes (e.g., between close air support [CAS] and surveillance missions), and close personal relations between forces in the field.[4]

Unfortunately, these wartime adaptations are unlikely to be maintained unless more formal measures are taken to legitimate them and make them permanent. USAF doctrine, concepts of operation, and tactics, techniques, and procedures (TTPs) must be more *proactively* adapted to distributed, ground-centric operations in Iraq and Afghanistan. USAF assets must be *responsive* to ground commanders' needs in Iraq and Afghanistan and be as *effective* as possible in providing that support. There is also increasing concern over *efficiency*—e.g., that expensive USAF platforms such as the F-16 are being used in CAS and nontraditional ISR (NTISR) missions (thereby "burning out" aircraft much faster than intended, with severe consequences in future years) that perhaps could be conducted better (more efficiently and with less cost) by other types of manned or unmanned platforms (e.g., Predator, new counterinsurgency aircraft).

Additionally, in Iraq and Afghanistan, as well as around the world, there is increasing demand for advisory assistance to help build partner capacity to defeat terrorist and insurgent groups and extend governance to populations vulnerable to extremist ideology and infil-

---

[4] Col Curtis Neal (USAF, Ret.) notes that during the Cold War, the Air Support Operations Center (ASOC) and Tactical Air Control Party (TACP) functioned quite well with just rated fighter/bomber Air Liaison Officers (ALOs) and 1C4 (formerly Tactical Air Command and Control Specialists). Today's environment, for both major combat operations and IW, requires not only a mix of USAF Air Force Specialty Codes (AFSCs), but also better joint command and control integration with the Army, SOF, and other operators in the battlespace. This is central to the Joint Air Ground Integration Cell (JAGIC) concept.

tration. The USAF must enhance and expand its capacity to build lasting relationships with willing and competent partner nations (particularly their air units), to train and advise those partners, and to provide equipment suited to those nations' milieus.

Air Operations in Iraq

The Air Force has been conducting operations in and around Iraq continuously for nearly two decades, and in and around Afghanistan for much of the current decade. Iraq and Afghanistan have become yardsticks of the USAF commitment to irregular warfare, both in perception and reality despite the fact, as discussed below, that Iraq and Afghanistan are not representative of most future U.S. IW operations. Prudence, however, dictates that the Air Force should plan to continue heavy involvement in direct and indirect operations in these two countries for some years to come.

The United States and Iraq have signed a Security Agreement whereby U.S. forces must withdraw from Iraq by December 31, 2011.[5] Nobody knows for certain what will happen between now (fall 2009) and then. The Iraqi government could declare that all foreign forces in any form must leave as dictated in the Security Agreement. In this case, the Air Force may need to maintain a robust posture nearby in case the United States needs to quickly surge to reinforce Iraqi forces or conduct other urgent operations in Iraq. On the other hand, the Iraqi leadership could ask U.S. forces to remain with a smaller, less visible presence that focuses on continuing to build Iraqi security forces while providing niche capabilities in support of Iraqi operations. This would require a renegotiation of the U.S.-Iraq Security Agreement. In this case, the Air Force may be left with ongoing, indefinite-duration missions after significant withdrawals of ground forces.

It would be dangerous to assume a best-case scenario in which demands on the USAF in Iraq simply vanish after the U.S. planned

---

[5] "Agreement Between the United States of America and the Republic of Iraq on the Withdrawal of United States Forces from Iraq and the Organization of Their Activities During Their Temporary Presence in Iraq," November 17, 2008.

withdrawal at the end of 2011.[6] Moreover, if the USAF were tasked to continue operations in Iraq beyond 2011, it could not wait to begin preparing *after* the fact—it would need to begin now. Thus, for planning purposes the USAF should prepare to keep sufficient assets in-country that may be needed to protect U.S. forces and bases, to support Iraqi ground forces, to protect Iraq's air sovereignty, and to train and advise Iraq's nascent air force. The transition from Coalition-led to Iraqi-led operations would transform USAF operational requirements, concepts of operation, and TTPs, especially in providing CAS, armed overwatch, and ISR for Iraqi forces, which are light-infantry oriented, currently with little or no organic heavy weapons such as mortars and artillery. And it will be years—well beyond the point that most Coalition ground forces have largely withdrawn—before the Iraqi Air Force will be able to provide significant support to their ground forces or to ensure air sovereignty. The USAF may have to fill this gap while continuing its efforts to build the Iraqi Air Force.[7]

Air Operations in Afghanistan and Beyond

Given the worsening conditions in Afghanistan, demands there for U.S. airpower will rise considerably. The lack of roads, the rugged terrain, and the distances between distributed battle areas will increasingly highlight the flexibility and reach of airpower. In addition, the safe haven that al-Qaeda and the Taliban enjoy in Pakistan greatly facili-

---

[6] While the future course in Iraq is uncertain, our suggested initiatives are not sensitive to these uncertainties, but rather provide hedging options and enduring IW value across the spectrum of possible long-term Air Force missions in Iraq (and elsewhere) following the June 30 pullout from Iraqi cities and the "complete" pullout of Coalition forces in 2011.

[7] The commander of Multi-National Forces in Iraq, General Ray Odierno, commented in early 2008 that U.S. air capabilities could be needed in Iraq for another five to ten years. More recently, Iraqi Prime Minister Nuri al-Maliki left the door open to continuing U.S. military presence after 2011: "If Iraqi forces need more training and support, we will reexamine the [status of forces] agreement at that time, based on our own national needs." See Department of Defense, "DoD News Briefing with Lt. Gen. Odierno from Iraq," January 17, 2008; "A Conversation with Iraq's Prime Minister Nuri al-Maliki," United States Institute of Peace, July 23, 2009; and David E. Thaler et al., *Future U.S. Security Relationships with Iraq and Afghanistan: U.S. Air Force Roles*, Santa Monica, Calif.: RAND Corporation, MG-681-AF, 2008.

tates their operations across the border into Afghanistan. This provides an impetus for the United States to work with the Pakistani government to apply airpower to best effect in an effort to eliminate extremist strongholds in the frontier areas, to patrol the borders, and to extend governance there. In addition, recent decisions to greatly expand the size of the Afghan National Army presage a long-term requirement for the USAF to help build the Afghan Air Corps and support the extension of Afghan government reach into rural areas.

Despite the criticality of operations in Iraq and Afghanistan, however, they are by no means archetypal scenarios for IW operations. Like the other services, the Air Force must prepare to work by, with, and through partners with diverse needs and at varied developmental levels by helping build their capabilities and capacity, even as the Air Force works to enhance its contribution to discreet U.S. direct operations against terrorist groups. This global demand for the expertise of U.S. airmen is neither static nor well defined, as yet, but its magnitude is widely recognized as being well beyond current U.S. capacity.[8]

Adapting the U.S. Air Force Approach to Irregular Warfare

Any Air Force commitment to IW must be in addition to ensuring that power projection for deterring regional or near-peer adversaries and, should deterrence fail, for fighting large-scale conflicts, including the need to address nuclear threats to the United States and its allies, is not compromised. In this balancing act it must be recognized that although IW and traditional warfare operations and systems may share much in common, IW operations are not a lesser included case of major combat operations. These differences are particularly apparent when one recognizes that IW is more about influencing relevant populations and operating by, with, and through partner nations, and that military force is

---

[8] See Adam Grissom and David Ochmanek, *Train, Equip, Advise, Assist: The USAF and the Indirect Approach to Countering Terrorist Groups Abroad,* Santa Monica, Calif.: RAND Corporation, MG-699-AF, May 2008, not available to the general public.

secondary to other instruments of national power (economic and political) in most IW contexts. Fortunately, there appears to be enough commonality between IW and conventional operations that fairly modest resource shifts and some new investments can "balance the force" appropriately across irregular and traditional warfare mission needs.

Solutions for addressing IW issues would start at the highest levels of the USAF with the recognition of the need for direction and discipline to develop and implement initiatives to reshape and rebalance the force.[9] From wide-ranging discussions the study team has had across the Air Staff and the major commands (MAJCOMs), we believe that this would require a cultural adaptation in the Air Force as an institution. To achieve this, the Air Force would need to adopt an incremental approach by augmenting selected elements and resisting the temptation to focus only on platform solutions. This approach should enable Air Force leaders to grow capabilities as demand dictates. Many relevant needs have been identified from analyses conducted by the Office of Air Force Lessons Learned, such as the comprehensive and insightful assessments of current operations in Iraq and Afghanistan as well as other engagements.[10] While the lessons are being "learned," the analyses that turn lessons learned into actions implemented seem to be lagging.[11]

---

[9] It may seem at first blush that within the USAF, IW is an "AFSOC mission" and that, consequently, IW enhancements are an AFSOC responsibility. While AFSOC has important roles in IW, rebalancing the force is a servicewide responsibility. IW is a total force issue requiring a total force mindset and culture change.

[10] See Office of Air Force Lessons Learned (HQ USAF/A9L), *Airpower in Irregular Warfare*, Washington, D.C., September 25, 2008; and Office of Air Force Lessons Learned (HQ USAF/A9L), *Integration of Airpower in Operational Level Planning Report*, Washington, D.C., August 22, 2008.

[11] It is worth noting that the Army has a process for analyzing its findings through the Combined Arms Center (CAC) at Fort Leavenworth. Along with the TRADOC Analysis Center (TRAC), the Center for Army Lessons Learned produces many excellent studies, and since CAC also "owns" the Combined Arms Doctrine Directorate and CAC training, which directs the Battle Command Training Program, they create a tight lash-up between lessons learned and implementation in doctrine and training. For over two years, the CAC has been sending trained analysts downrange to serve in the maneuver unit headquarters to assist in identification of lessons and analysis of operations.

The good news is that needed tactical-level changes are bubbling up from operators in the field (e.g., as documented by the aforementioned Office of Air Force Lessons Learned and the efforts of other groups such as the Headquarters Air Force IW Task Force and IW Tiger Team, in addition to advice provided by AFSOC based on its long experience in the IW realm). There is inherent (but in some cases unrealized) flexibility in the USAF Total Force to meet these challenges. Operational-level changes, however, are more challenging—and it is here where senior leadership direction is most needed. Many of the near-term solution vectors we identify focus on the development of more specialized human capital and processes. We feel that these nonmaterial career management, education and training, and command and control processes can and should be acted on immediately, both to help in reshaping current operations in Iraq and Afghanistan (and other places) as well as to begin adapting USAF culture to make it more sensitive and responsive to emerging IW needs such as building partnerships and partner capacity.

Finally, there are material issues that need to be addressed, some of which are under way (e.g., the MC-12 Project Liberty, a program to meet near-term shortfalls in ISR coverage with manned platforms). Others are more long term in relation to the current operations in Iraq, Afghanistan, and elsewhere, and may have a relatively significant effect on force structure over the next couple of decades.

CHAPTER THREE

Exploring Courses of Action for Irregular Warfare

In the previous chapter we suggest a number of current shortfalls and anticipated future needs that the U.S. Air Force could address. In this chapter we identify four "dimensions" in which the Air Force could act to expand its contributions to joint IW operations, both direct and indirect. They are (1) adapting the Air Force as an institution to embrace an IW mindset as it maintains the capability to deter and, if necessary, to defeat regional adversaries; (2) enhancing its capacity quickly to meet the pressing needs of commanders in Iraq and Afghanistan; (3) ensuring that it can meet global demands to support partner nations against extremist threats to their stability and to the well-being of their populations; and (4) building capabilities and capacities to address future challenges in the long war.

Based on those four dimensions, we developed four macro-level courses of action, each of which addresses USAF actions in a given dimension. We formulated the CoAs using an incremental approach. Rather than treating the CoAs as separate alternatives from which decisionmakers select a single CoA, each CoA builds upon the previous one. A CoA contains a bundle of individual initiatives related to the organizing theme (e.g., institutional commitment).

The CoAs are presented in order of their urgency or time-horizon, specificity, scope, and increasing resource requirements:

- **CoA 0: Set the Climate.** This CoA is foundational to all other CoAs. It seeks to deepen the institutional commitment to developing IW expertise and providing relevant capabilities. Pursuing a set of initiatives such as those in this CoA is not optional, but essential, if the Air Force is to truly embrace irregular warfare as a USAF competency. It is for this reason that we call it "CoA 0."

- **CoA 1: Succeed in Iraq and Afghanistan.** This CoA undertakes new efforts to enhance the effectiveness and efficiency of USAF operations in Iraq and Afghanistan. The intent is to show immediate effects in both fights in the near and middle terms, enabling airmen to play a greater role in strategy, planning, and operations. Early efforts should focus on ensuring that commanders and forces in both countries have what they need.
- **CoA 2: Support Partners Globally.** This CoA emphasizes meeting the demands emerging from the rest of the long war against both violent global extremists and local insurgents. The focus here is on enhancing USAF capabilities for building partner capacity and for working by, with, and through partners to generate significant results. The CoA is based on initiatives to grow airmen who can train, advise, and assist their counterparts in partner nations to pressure terrorist networks.
- **CoA 3: Ensure Future Access and Action.** This CoA takes a longer-term focus on rebalancing the force. It provides the preponderance of resources for material (e.g., platform) initiatives.

In many cases, suggestions by the Air Force itself were the source of the initiatives that we describe in these CoAs. In fact, most of the initiatives identified here are already being considered and, in some instances, pursued by Air Force organizations.

Effectively rebalancing Air Force capabilities to incorporate irregular warfare will likely require action in *all* CoAs. Prioritizing and timing these efforts involves a series of decisions (and, in some cases, more studies). We have adopted an incremental crawl-walk-run approach, but it is possible that the leadership may wish to treat these options as a "Chinese menu" from which it selects various combinations of initiatives. Decisionmakers can consider initiatives individually on their merits. For the most part these initiatives are related but not strongly interdependent, although there are subsets of initiatives within each CoA that should probably be considered as a unit—they make the most sense as parallel, mutually supporting efforts.

The costs (material and nonmaterial) of these initiatives have been roughly estimated, with costs and uncertainty increasing as one progresses from the first to the last CoA. However, the specific details of the initiatives are less important at this point than the overall paradigm we have attempted to nurture. Numerous related or supporting studies and proposals across the USAF can supply the details once a general set of CoAs or vectors is chosen.

We describe the four courses of action in detail in the following sections and offer rough estimates of the resources that might be needed to implement a course of action. Descriptions and estimated costs of each initiative may be found in Appendix A (references to these initiative descriptions appear in parentheses after we introduce each initiative below). Appendix B provides a consolidated estimate of resources that would be associated with the CoAs through the Future Years Defense Plan and out to 2030.

Course of Action 0: Set the Climate

To ensure that individual initiatives gain traction within the Air Force, the leadership needs to act to set the conditions for the Air Force's institutional culture to value and nurture the initiatives. To this end, we identify initiatives emphasizing leadership, career management, education and training, and concept development.

Provide Leadership Emphasis. There can be no substitute for frequent emphasis by the Air Force's top leadership on IW as an institutional priority, in CORONA (USAF four-star level) meetings, in communications across the USAF as a whole, and in interactions with DoD, interagency partners, and Congress. The leadership would reinforce this emphasis with concrete actions affecting careers, organizations, and funding. (See Appendix A, p. 47.)

Designate Headquarters Air Force Integrator for Irregular Warfare. CoA 0 envisions the creation of a permanent, high-level management organization led by a Headquarters Air Force general officer to monitor and direct IW activities in the Air Force and to ensure unity of purpose. It would serve as the focal point and integrator for the entire Air Force on IW-related issues, would represent the USAF in joint and interagency forums, and would advocate for resources within the Air Force corporate structure. Its charter would include overseeing the development of IW strategy and doctrine; fostering IW-related experience in career management; conducting and reviewing studies and analyses; ensuring that IW is properly represented in training syllabi and designed operational capability statements; and taking the lead in IW-related interactions with other services—especially the U.S. Army—and joint and interagency organizations, such as the U.S. Special Operations Command (USSOCOM), U.S. Joint Forces Command (USJFCOM), and non-DoD government agencies.

We identify three options for such an organization at Headquarters Air Force. The first would make the current IW Task Force a permanent organization under a two-star reporting to a three-star deputy chief of staff. The second option would be a two-star with visibility across all functional areas reporting directly to the Vice Chief of Staff. The third option would do the same, albeit through the creation of a Deputy Chief of Staff for IW organization (perhaps "A11" in Air Staff parlance) led by a two- or three-star. (See Appendix A, p. 48.)

Emphasize Irregular Warfare Career Incentives. The third initiative in CoA 0 is a broad realignment of career management that encourages and rewards experience in irregular warfare—especially advisory skills and regional, cultural, and/or linguistic expertise.[1] Based on discussions with a wide range of airmen, such experience is currently perceived by many as not conducive to career advancement. A goal of this initiative is to help change this perception throughout the force. Besides providing incentives and rewards, the USAF would emphasize selecting the right personnel as advisors and for other specialized duties and then tracking them throughout their careers to ensure that

---

[1] A standard solution advocated by some quarters of the Air Force in relation to training, advising, and assisting foreign partners has been that "everybody will learn English," and then partners can learn how to build and operate an air force. Indeed, English is the International Civil Aviation Organization language, but that does not mean that all maintainers, logisticians, recruiters, etc., should be expected to speak English. Clearly, improved foreign language skills in the USAF are a critical prerequisite to more effective and efficient indirect support of partner nations.

the USAF can present a flexible, "scalable" IW capability to combatant commanders.[2] Besides tracking special experience identifiers (SEIs) and developing criteria for personnel selection for specialized duties, this initiative would involve instructing promotion boards to emphasize IW experience, establishing new IW-related career requirements, and placing a high priority on rapidly assigning appropriate personnel to IW-related manpower slots, including joint and interagency billets. Total Force options could be actively considered to enhance the experience base of the Air Force.[3] (See Appendix A, p. 50.)

Expand Irregular Warfare Curricula in Education and Training. Next, the Air Force would increase the exposure of its airmen to IW concepts in education and training throughout their careers. IW curricula could be integrated beginning with accession, Reserve Officer Training Corps, Officer Training School, U.S. Air Force Academy, and enlisted training. It would also extend to professional military education, which would include emphasis on IW campaign planning to prepare USAF planners to work with joint and interagency partners to bring airpower most effectively and efficiently to bear in IW fights and activities.[4] (See Appendix A, p. 51.)

Expand Coalition Irregular Warfare Center. The final initiative in CoA 0 would expand the responsibilities of the Coalition Irregular Warfare Center (CIWC) to solidify an Air Force focal point for IW-related concept development, technology exploitation, and strategic assessment. Such an IW center (whether an expansion of the existing

---

[2] DoD guidance calls for the services to "maintain scalable organizations to train and advise foreign security forces and security institutions (unilaterally or as part of civilian-military teams) in permissive and uncertain environments." DoDD 3000.07, p. 8. We interpret "scalable" to mean that the capacity for such activities can be adjusted on demand.

[3] The USAF should also consider the need for IW-related AFSCs to help track IW-qualified personnel and to help manage career development and rewards.

[4] These efforts should lead the USAF toward greater participation in combined-arms operational and tactical training exercises and experimentation using federated simulations and/or live-fire venues. A good example of such an emerging venue is the Fort Sill Fires Battle Lab series of experiments known as Earth, Wind, and Fire, which looks at integration of airspace control and fires. The USAF has informally supported this experiment for the past three years.

CIWC or a new organization) would identify and evaluate potential operational uses of emerging technologies; develop IW-specific TTPs; coordinate with research, development, test, and evaluation (RDT&E) organizations such as the Air Force Research Laboratory, Defense Advanced Research Projects Agency, and Department of Energy labs; and address seams in interoperability and transferability between the USAF and partner air arms through strategic assessments of partner needs. In addition, this initiative calls for the creation of small detachments of IW center technologists embedded in key organizations—air components, MAJCOMs, and intelligence agencies—to help anticipate operational needs and drive formulation of solutions. (See Appendix A, p. 52.)

Estimated resources that would be required to implement these initiatives would include 181 personnel, an initial investment of about $7 million (mainly for office equipment), and an annual operations and support (O&S) cost of about $30 million.

Course of Action 1: Succeed in Iraq and Afghanistan

Course of Action 1 would boost Air Force capabilities for the planning and conduct of operations in Iraq and Afghanistan by pushing more USAF capability and capacity forward, preparing the USAF for what could be a long stay and a more central role (if U.S. and Iraqi leaders agree). It would also augment ongoing efforts to establish the Iraqi Air Force and Afghan Air Corps as strong, independent providers of indigenous air power—goals that will allow the Air Force to gracefully disengage from direct support obligations in Iraq and Afghanistan. CoA 1 initiatives are designed to improve the effectiveness and efficiency of air support to distributed, decentralized ground operations in the two fights and to position the USAF to support local partners as U.S. forces draw down. They would increase the USAF's focus on the operational level of IW, helping to adapt processes for IW requirements. They would integrate CAS, lift, ISR, medical evacuation, and information operations, and designate senior IW-experienced airmen to work with counterpart U.S. Army (and eventually Iraqi and Afghan) planners.[5] Implicit in these initiatives is the goal of strengthening bridges to U.S. ground components at the operational level and clearly demonstrating an Air Force commitment to supporting joint IW operations—many of which are ground-centric—over the long haul.

Push More Air Force Capability and Capacity Forward to/Within the Theater

Embed Air Expertise Forward. The first initiative focuses on improving the command and control shortfalls in the theater to promote more effective integration of airpower with ground operations. The initiative calls for a senior air commander with planning and tasking authority (and staffs) to be assigned to Baghdad and Kabul to work with the ground commander to plan and implement full air-ground integration across the theater of operations. It also calls on the Air Force to embed a broad range of air expertise at lower levels of U.S. Army organizations (especially at division and brigade level, but conceivably at battalion and below) to facilitate the best application of airpower in planning and execution of specific air-ground operations. Expertise would also extend to assessment of potential collateral damage resulting from the use of airpower—a critical element of IW that often has dire strategic effects on the success of U.S. and partner efforts to insulate and protect local populations from terrorists and insurgents. Embedded air cells would include ISR, electronic warfare, information operations, cyber operations, space, mobility, and logistics expertise to augment tactical air control parties at some 24 brigade combat teams in Iraq and Afghanistan.[6] (See Appendix A, p. 54.)

---

[5] As stated previously, initiatives developed here and in other CoAs are based on the collective body of analysis and operational experience we accessed for this quick-turn effort. Additional force-effectiveness and cost analyses may be required to implement the initiatives.

[6] This initiative merely scratches the surface of ongoing USAF efforts to adapt the major combat operation–centered theater air control system to the exigencies of operations in Iraq and Afghanistan. For a source of recommendations along these lines, see Col Gary Crowder, "Air-Ground Integration in the 21st Century," documented briefing, U.S. Air Forces Central Combined Air Operations Center, Al-Udeid Air Base, Qatar, April 2008. One suggestion is to bring back the Battalion Air Liaison Officer. From Corps down to Battalion headquarters,

Add Joint Terminal Attack Controllers. The number of available Joint Terminal Attack Controllers (JTACs) would be increased to ensure a USAF ability to provide CAS and armed overwatch to U.S. and Iraqi ground forces during and after drawdown of U.S. brigade combat teams. This would reduce deployment stress on JTACs and personnel in related Air Force career fields and ensure that tactical air control parties can control CAS throughout Iraq even as brigade combat teams bring their JTACs home with them. Equipping some units with a dedicated CAS platform for counterinsurgency (COIN) (see below) would support the increased requirements for live CAS training of Air Force JTACs and their Army counterparts. (See Appendix A, p. 56.)

Push and Sustain Airborne ISR and PED. The next initiative builds on "Project Liberty," the ongoing effort to deploy 37 MC-12 ISR platforms to augment such unmanned systems as Predator, which are being increased to meet expanding demand from ground commanders. This CoA 1 initiative would add 30 MC-12s and processing, exploitation, and dissemination (PED) capability to provide a more robust rotation base and allow necessary continuation training that is lacking in the existing program. The manpower and other resources associated with this initiative would be in addition to resources allocated to Project Liberty. (See Appendix A, p. 58.)

Emphasize Innovative Information Operations. The Air Force should also consider pushing forward a range of information operations capabilities, including cyber operators. By putting information operations expertise in the combined air operations center (CAOC) and filling USAF billets in other organizations such as the National Security Agency, this initiative would enable U.S. commanders in Iraq and Afghanistan to better exploit USAF technical capabilities in computer network operations, electronic warfare, and multiple intelligence sources in the fight against terrorist and insurgent networks. These capabilities would help counter adversary use of the Internet, exploit technology for directed psychological operations, and provide timely electronic warfare support to ground forces. (See Appendix A, p. 60.)

---

these personnel need to be part of the existing aligned TACP, providing cross-functional expertise, but speaking with a single Air Force voice.

Provide Transferable, Counterinsurgency-Dedicated Close Air Support and Armed Overwatch Platform. The USAF is considering fielding a to-be-determined OA-X counterinsurgency-dedicated CAS platform that would provide light attack and ISR capability in support of U.S. and partner ground forces, but that could also be transferred as needed to the Iraqi and Afghan air arms. Such a dedicated counterinsurgency platform would help ensure persistent presence and engagement with Iraqi and Afghan partners while lowering operating costs and reducing the excessive flying-hour demands for high-performance aircraft such as the F-16. Transferability would depend on the platform's flexibility, sustainability, ease of maintenance, and interoperability with U.S. and Coalition forces. It is commonly held that partners are more likely to want aircraft that U.S. forces are flying to great effect; such a platform could be responsive and effective in support of ground forces engaged in IW while whetting the appetite of partners who are prematurely looking to acquire high-performance jet aircraft such as the F-16. For illustrative purposes, this CoA 1 initiative calls for a wing of about 100 OA-X light attack aircraft. (See Appendix A, p. 71.)

Develop and Procure Light Cargo Aircraft. Similarly, the Air Force could consider operating a transferable light cargo aircraft (aircraft in the 3,000- to 6,000-pound payload range) to help Iraq and Afghanistan reinforce and extend governance to remote and/or undergoverned regions with locally unobtrusive platforms (the more unobtrusive U.S. presence becomes, the better for the legitimacy of the local government in the eyes of the population).[7] These aircraft could support distributed U.S. and partner military operations, including resupply and medical evacuation. This initiative would provide the beginnings of a "Small Multi-Mission Air Fleet" with 30 aircraft patterned after AFSOC's Non-Standard Aviation program (a family of regionally suitable and unobtrusive aircraft). (See Appendix A, p. 72.)

---

[7] PAF work on intratheater airlift has found that although the C-130 is the most cost-effective platform for meeting the theater lift requirements of major combat operations, smaller cargo aircraft—in the Cessna Cargomaster, Beachcraft C-12, and EADS CASA C-212 class—are the most cost-effective for smaller-payload missions such as supporting special operations forces.

Prepare the Air Force for Protracted Joint Operations in Iraq and Afghanistan

We offer three additional initiatives to help position the Air Force for protracted operations in Iraq and Afghanistan and for a larger role in planning and execution of operations.

Synchronize Strategic Communications. First, the USAF should strive to better synchronize its strategic communications efforts with national, combatant command (COCOM), and even local government messages through changes in emphases and processes. The primary audience should be the local population, not U.S. public opinion. USAF communications would advise Iraqis and Afghans of the nature and purpose of air operations—to support and legitimize their governments, defend them and their security forces from adversaries, deliver humanitarian aid, and build infrastructure. Strategic communications must be better integrated with operations, including preparing messages before operations take place and countering adversary propaganda on the operations, especially if they result in collateral damage. No new systems or billets are required for this initiative, but the U.S. Air Forces Central commander could direct changes in Air Force communications to external audiences in Iraq and Afghanistan. (See Appendix A, p. 61.)

Mitigate Agile Combat Support Shortfalls. Second, shortfalls in key agile combat support (ACS) personnel—security forces, logistics personnel, civil engineers, medical personnel, and others—should be addressed to ensure their long-term ability to execute their missions in Iraq and Afghanistan. These forces are at the center of supporting U.S. direct operations as well as providing services to, and enhancing the well-being of, local populations. Their activities are critical for insulating residents from terrorist and insurgent influence and legitimizing local and central governments. This initiative would reduce stress in key career fields affected by both USAF and "joint expeditionary" tasking by providing 950 additional personnel.[8] These additional

[8] Joint expeditionary tasking, a USAF term formerly called "in-lieu-of" tasking, refers to requests from joint commanders for airmen to perform missions that are outside their normal competencies or that are in place of military members from other services. Based on RAND analysis using such requests as a baseline, we identified two options for the combat support initiative. One option assumes a reduction in security services at home station of 20 percent,

personnel would be needed in anticipation of changes in demand in Iraq and Afghanistan over the next few years. In Iraq, a withdrawal of most U.S. ground forces could serve to reduce demand from joint tasking; on the other hand, continued Air Force presence there with fewer ground elements might actually increase ACS demands (especially in force protection, building infrastructure, and providing medical support). In Afghanistan, demands on the Air Force can be expected to rise across the board. (See Appendix A, p. 62.)

Operationalize Analyses. A third and related initiative calls for capabilities to ensure that lessons *observed* in Iraq and Afghanistan become lessons *learned and acted on* by instituting analytic capabilities to more readily and quickly incorporate lessons into force planning, TTPs, and air-ground operations. This would involve developing analytic frameworks and processes, and providing appropriate manning, to focus on operational-level measures of effectiveness, improve data collection, and provide the means to act quickly on findings. This initiative would stand up a dedicated air-ground analysis cell in the CAOC with reachback support from Army and Air Force analysis cells in the continental United States. (See Appendix A, p. 63.)

Broaden, Formalize, and Accelerate Efforts to Build Iraqi and Afghan Air Arms

Finally, we suggest three initiatives to broaden and formalize efforts to build the Iraqi and Afghan air arms.

Expand Advisor Training. First, a tailorable, institutionalized, well-resourced GPF advisor training school should be established in Air Education and Training Command (AETC) that adapts the number of training days, curriculum, and training equipment (including partner aircraft) to the task and cultural requirements that advisors will encounter in partner nations. This initiative would institute more

---

plus an influx of contract guards who cover an additional 20 percent of home-station security requirements. A second option removes these assumptions. Both options assume full reserve component mobilization and a rotation base supporting a 2:1 dwell. The first option—the one we use above—requires a manpower increase of 950 but might threaten the long-term viability of several career fields, especially security forces; the second option would require an additional 4,125 personnel for a total of 5,075 spread across several career fields.

robust cultural and language skills training beyond "familiarization," ensure that advisors are identified and tracked (a corollary to the CoA 0 career management initiative), weed out individuals who are unsuited to advising, collect lessons from the field to improve curricula, and ensure that advisors are properly equipped for downrange deployment. The pipeline would sustain a force of 800 GPF advisors for Iraq and Afghanistan, plus about 400 AFSOC combat aviation advisors, as called for in the following initiative.[9] (See Appendix A, p. 64.)

Add Combat Aviation Advisors. The next initiative would establish a sister squadron of combat aviation advisors for the existing 6th Special Operations Squadron in AFSOC. The purpose of this initiative is twofold. First, as the Iraqi Air Force and Afghan Air Corps continue to expand, there will be an increasing need for advanced tactical training that enables Iraqi and Afghan rotary- and fixed-wing aircrews to conduct sophisticated air-ground operations. This is an area in which combat aviation advisors excel and to which they bring specialized training skills. Thus, the demand for combat aviation advisors can be expected to grow. Second, a sister squadron would position AFSOC to implement an emerging institutional concept for aviation advisors developed by USSOCOM to meet a DoD directive that designates the command as the lead security forces assistance proponent. The new squadron would include additional partner aircraft for training and currency of the squadron's advisors. (See Appendix A, p. 68.)

Build Regional Air Academies. The final initiative in CoA 1 would build a regional air academy focused on U.S. Central Command (USCENTCOM), in general, and Iraq and Afghanistan, in particular. Patterned after the Inter-American Air Force Academy that provides education and training for rated officers from Central and South American countries under U.S. Southern Command, the USCENTCOM academy would be an international military education and training institution providing training in instrument and other flight activities,

---

9 The concept here would be to train all student advisors up to the same GPF standard, but to identify a subset of students who could be candidates as air commandos in AFSOC. These candidates would then go through further training at AFSOC to meet special operations standards.

maintenance, airbase defense, logistics, and other courses for some 500 students primarily from middle ranks. The purpose would be to meet future demand for individual military education and training for Iraqi, Afghan, and other airmen; enhance and sustain USAF investments in tactical training for those airmen; establish enduring U.S. relationships with future leaders; and foster cooperative relationships among airmen from multiple regional countries. (See Appendix A, p. 70.)

Implementing the initiatives in CoA 1 would require an estimated 3,600 personnel, 200 aircraft, about $1.9 billion in initial investment (mainly for aircraft), and some $423 million in yearly O&S costs. The MC-12, OA-X, light cargo aircraft, and agile combat support initiatives make up a preponderance of these resource requirements.

Course of Action 2: Support Partners Globally

Course of Action 2 addresses IW demands in the rest of the world, especially those demands related to building partner capacity and working by, with, and through those partners, even as efforts to expand the Iraqi Air Force and Afghan Air Corps continue at high levels. As noted previously, a robust effort to improve the competency and professionalism of partner militaries will be essential if the United States is to place terrorist networks under constant pressure worldwide. Long-term relationships to assess, train, equip, advise, and provide operational assistance to partners can also be invaluable as sources of situational awareness and influence, which the United States can exploit to gain access in time of need. In addition, this CoA would enable USSOCOM to leverage Air Force capabilities for joint and interagency requirements.

Stand Up General-Purpose Forces Advisor Unit and Elements; Add Combat Aviation Advisors. The first three initiatives in CoA 2 would establish one advisory wing in each of the general-purpose and special operations forces and broaden advisor training to sustain this expanded cadre of USAF advisors.[10] The two wings would bring USAF

---

[10] For information on IW wing and advisory concepts, see "USAF Irregular Warfare Concept," Air Force Special Operations Command White Paper, May 2007; and Alan Vick et al.,

steady-state advisory capacity outside Iraq and Afghanistan from the current 20–24 military training team–like missions per year to 130–140 per year, equivalent to estimated demand.[11] The purpose of the IW wings would be to assess, train, advise, and assist partner nation air arms in building institutional capacity and operational capability. When necessary, this could include operational assistance with mobility, ISR, and potentially strike missions during partner nation operations. The wings would enable the Air Force to support a *persistent* advisory presence in partner nations to help ensure retention of imparted skills, continuous improvement in capabilities and capacity, and maintenance of relationships and access. The wings also would provide a dedicated rotation base for advisors, continuation training, repositories for lessons learned (serving as the "institutional memory"), and opportunities for airmen who desire a career as an advisor. Finally, they would provide the USAF with a scalable, tailorable advisory capability with reachback into the rest of the force. The wings would support implementation of USSOCOM's concept for security forces assistance and would be manned by advisors (nominally, 1,000 in the GPF wing and 500 in the air commando wing) from a wide range of career fields—pilots, maintainers, civil engineers, medical personnel, security forces, etc.

Each wing would contain five COCOM-aligned squadrons with suitable regionally oriented advisors and equipment, including equipment for improving both civilian and military air infrastructure. The division of labor between the two wings would follow the existing mix of skill and mission sets of the two communities. The general-purpose force wing would help partners expand and enhance their force structure (e.g., through basic flying training) and build institutional capacity (e.g., forming logistics systems, building training pipelines, etc.). It could also supplement contingency response groups. (See Appendix A, p. 66.) The AFSOC wing would provide advanced training and assistance to partners, extending existing capabilities for tactical employ-

---

Airpower in the New Counterinsurgency Era: The Strategic Importance of USAF Advisory and Assistance Missions, Santa Monica, Calif.: RAND Corporation, MG-509-AF, 2006.

11 See Grissom and Ochmanek, 2008, pp. 40–42.

ment and sustainment (e.g., use of night-vision goggles, integration with ground components for air assaults, etc.), especially in austere, less permissive environments. (See Appendix A, p. 66.) Despite some divergence in skill sets, however, the two wings could coordinate mission planning and work closely together when downrange.[12]

To support planning and assessment of demands for advisory activities, CoA 2 contains an initiative to assign airmen with advisory and regional expertise to the air components of each COCOM. These advisors would capture air-related demands of partner nations in theater engagement plans and would help formulate tasking to USJFCOM, USSOCOM, and ultimately, USAF advisory units. Their mission would be to remain in close contact with the air arms and governments of partner nations in the COCOM's area of responsibility and to ensure that U.S. and partner nation needs are being met in the air realm. There could be tight coordination with (and support from) the USAF advisory wings and such efforts as the State Partnership Program through the Air National Guard. (See Appendix A, p. 68.)

Expand Advisor Training. The AETC training pipeline would be expanded further under CoA 2 to sustain a steady-state force of an additional 1,500 advisors in the GPF and air commando wings.[13] This would include additional instructors, partner aircraft, and other equipment. (See Appendix A, p. 64.)

Mitigate Agile Combat Support Shortfalls. CoA 2 would include new concepts—such as ACS-centered mobile training teams and other employment concepts—by which agile combat support personnel are employed in support of efforts to train, advise, and assist partner nations. These concepts would rationalize presentation of ACS capabilities in support of building partner capacity, including base support,

---

[12] In practical terms, the division of labor between the two wings would not be this stark. For example, it may be appropriate for combat aviation advisors to help small, underdeveloped countries build institutional capacity. Likewise, GPF advisors could at times provide advanced training to partners. Skill and mission sets could be worked out by the Air Force over time.

[13] As noted earlier, AETC would train all advisors to GPF standards, while AFSOC would conduct the specialized training needed for special operators once an AETC graduate enters that community.

institution-building, humanitarian aid, and medical and construction influence operations to help extend local governance and services. (See Appendix A, p. 62.)

Build Regional Air Academies. CoA 2 also calls for the establishment of regional air academies focused on each of the remaining commands—U.S. Pacific Command, U.S. Africa Command, and U.S. European Command.[14] Each institute would cater to the needs of the particular regions and the nations within. For example, the academy for U.S. Africa Command would likely focus on foreign internal defense, airspace management, and civil-military relations, while the U.S. European Command academy might focus more on the basics of coalition IW operations with East European nations. (See Appendix A, p. 70.)

Develop and Procure Light Cargo Aircraft. CoA 2 assumes that the Air Force will continue to deploy unmanned aerial systems at the fastest rate possible to expand the capability for ISR and armed overwatch on a global scale. The initiative to deploy a larger number of transferable, multi-mission light cargo/utility aircraft would further address the need for ISR as well as mobility in IW operations and would support both U.S. and partner demands with unobtrusive platforms. The family of aircraft would be modular and could deploy as mobility, ISR, medical evacuation, and even strike platforms. With one operational squadron deployed in each of the five overseas COCOMs, the USAF could use these aircraft to support country teams and to help whet the appetites of partners for their own airpower capabilities. They could also be used for operational assistance of partner security forces and to help indigenous governments extend services to outlying regions. At the same time, these light cargo aircraft would help the Air Force keep pace with and support the ongoing expansion of partner-building capacity in the other services. (See Appendix A, p. 72.)

Expand Air Force Human Intelligence. The Air Force should institute processes to better exploit the human intelligence value of airmen on overseas assignments, pursuing the idea that "every airman is a

---

[14] These would be in addition to the existing U.S. Southern Command academy and the CoA 1 USCENTCOM academy.

sensor." This initiative would provide formal education and training for airmen with human intelligence opportunities beyond foreign area officers, including attachés and security assistance officers, COCOM theater engagement plan teams, advisors, and National Guardsmen on State Partnership Program exchanges. It would also create processes to fully exploit the intelligence with processing, exploitation, and dissemination capabilities. (See Appendix A, p. 69.)

Overall, CoA 2 involves about 3,000 personnel, 255 aircraft, an investment of $2.3 billion, and an annual O&S cost of $374 million.

Course of Action 3: Ensure Future Access and Action

Course of Action 3 is aimed at rebalancing force structure to support direct and indirect IW operations over the long term. Its main objectives are to ensure global access and action in the future through cooperative partnerships, presence, infrastructure, intelligence, and suitable forces; enable special operations forces to operate in less permissive environments during counterinsurgency and counterterrorism operations (at times clandestinely); and provide an Air Force surge capability for future IW operations.

While the conflicts in Iraq and Afghanistan should not be considered models for future IW operations, it would be prudent to prepare for the possibility that the United States may again lead a large counterinsurgency and/or stability operation in the future. Moreover, as the United States becomes more involved in foreign internal defense around the world, demands for air support from both U.S. and partner forces can be expected to increase. Finally, even as the United States emphasizes operating by, with, and through partners, there will still be a need for U.S. forces to conduct direct operations against nonstate adversaries. As these demands become better understood over time, two initiatives—both of which expand upon capabilities addressed in CoA 1—could help position the USAF to support surge operations and other potential increases in demand for ISR, armed overwatch, CAS, and building partner capacity in multiple areas of responsibility.

Push and Sustain Airborne ISR and PED. The first provides additional manned ISR capability to complement what we assume will be continued expansion of unmanned ISR systems. As a placeholder, this initiative would procure an additional 93 MC-12s along with appropriate PED capabilities. (See Appendix A, p. 58.)

Provide Transferable, Counterinsurgency-Dedicated Close Air Support and Armed Overwatch Platform. The second initiative would procure an additional 300 OA-X aircraft for global operations. (See Appendix A, p. 71.)

Before implementation, the USAF should undertake an analytic assessment of the need for, and nature of, the larger numbers of USAF-flown, COIN-dedicated platforms that these initiatives entail.

We estimate the resources needed for these two initiatives include about 4,400 personnel, $4.7 billion in investment, and about $600 million in annual O&S costs once fully implemented with 393 aircraft.

Develop and Deploy the Next-Generation Gunship. The final two initiatives offer concepts for enhanced special operations capabilities in the out-years. A low-observable, next-generation gunship concept would enable responsive, tailorable, precise, and sustained fires in support of special operations day and night in semipermissive environments (e.g., those with man-portable air defense systems or antiaircraft artillery threats). Options could include manned or unmanned platforms and standoff precision-guided munitions and ISR concepts. The gunship would have a large magazine and allow persistence over target areas. The gunship concept could be a variant of the next-generation bomber; it is *not* related to the AC-27 Stinger concept that was considered as a midterm complement to the current AC-130. The concept to be pursued and the numbers involved are yet to be determined; for purposes of this effort we drew upon the next-generation gunship analysis of alternatives that RAND conducted in 2004; this analysis offered two equal-cost concepts of 24 manned gunships (which we use here as a placeholder) or 77 unmanned platforms.[15] (See Appendix A, p. 74.)

---

[15] For an analysis of gunship options, see Richard M. Moore, John Stillion, et al., *Next Generation Gunship Analysis of Alternatives: Final Report—Volume I*, Santa Monica, Calif.: MR-

Develop and Deploy Next-Generation SOF Mobility Platform. A low-observable mobility platform would enable special operations forces to conduct clandestine infiltration, resupply, and exfiltration operations in denied or politically sensitive areas and over great distances. As with the gunship initiative, we assume a buy of 24 manned aircraft, although the exact nature of the concept is yet to be determined.[16] (See Appendix A, p. 75.)

We estimate resources for the 48 special operations aircraft of both types to be 1,320 personnel, $20.5 billion in investment, and $216 million in annual O&S costs. Most of the resources would come into play beyond 2020, although some RDT&E would appear late in the Future Years Defense Plan.

Summary of Courses of Action

Table 3.1 provides a summary of CoAs and initiatives, with the initiatives grouped by functional area: institutional, operational capability, building partner capacity, and future procurement. Details for the initiatives appear in Appendix A in the same order found here.

If an initiative in this table is identified previously as part of a course of action, it receives a checkmark under that CoA. However, initiatives may also play roles in follow-on CoAs. As such, if an initiative also supports the goals or emphases of other CoAs, it receives a plus sign under those CoAs. Thus, for example, emphasizing IW career incentives (and the other initiatives in CoA 0) not only helps the USAF deepen its institutional commitment to irregular warfare capabilities—the goal of CoA 0—but also is crucial to pursuing any of the other CoAs. Likewise, while Iraq and Afghanistan are the objects of the

---

1795-AF, 2004; not available to the general public. The design options and tradeoffs between manned or unmanned gunships may require updated analysis.

[16] AFSOC commissioned Jacobs Sverdrup Technology, Inc., to perform an analysis of alternatives for a future mobility platform. See U.S. Air Force, *Advanced Special Operations Air Mobility Platform (M-X) Analysis of Alternatives*, September 30, 2004; not available to the general public.

Table 3.1
Summary of Initiatives to Support Courses of Action

Initiatives	CoA 0 Set the Climate	CoA 1 Succeed in Iraq and Afghanistan	CoA 2 Support Partners Globally	CoA 3 Ensure Future Access and Action
Institutional				
Provide leadership emphasis	√	+	+	+
Designate HAF integrator for IW	√	+	+	+
Emphasize IW career incentives	√	+	+	+
Expand IW curricula in education and training	√	+	+	+
Expand CIWC	√	+	+	+
Operational Capability				
Embed air expertise forward		√		
Add JTACs		√		
Push and sustain airborne ISR and PED		√	+	√
Emphasize innovative information operations		√	+	+
Synchronize strategic communications		√		
Mitigate agile combat support shortfalls		√	√	+
Operationalize analyses		√	+	+
Building Partner Capacity				
Expand advisor training		√	√	+
Stand up GPF advisor unit and elements			√	+
Add combat aviation advisors		√	√	+
Expand USAF human intelligence			√	+
Build regional air academies		√	√	+
Future Procurement				
Provide transferable, COIN-dedicated CAS and armed overwatch platform		√	+	√
Develop and procure light cargo aircraft		√	√	+
Develop and deploy next-generation gunship				√

Table 3.1 (continued)

Initiatives	CoA 0 Set the Climate	CoA 1 Succeed in Iraq and Afghanistan	CoA 2 Support Partners Globally	CoA 3 Ensure Future Access and Action
Develop and deploy next-generation SOF mobility platform				√
Total Estimated Resources				
Manpower	181	3,594	2,995	4,395 [1,320]
Total aircraft inventory	0	200	255	393 [48]
Investment	$7 million	$1.9 billion	$2.3 billion	$4.7 billion [$20.5 billion]
Annual O&S	$30 million	$423 million	$374 million	$598 million [$216 million]

initiative to provide 100 transferable, counterinsurgency-dedicated CAS platforms (OA-X), these platforms might also be used beyond those conflicts throughout the world (and their use in these two fights may also provide lessons for OA-X operations elsewhere).

On the other hand, we consider some initiatives to respond mainly to the needs of the CoA with which they are primarily identified. Embedding air expertise forward and adding Joint Terminal Attack Controllers are initiatives conceived as mitigating specific problems in Iraq and Afghanistan. Therefore, a checkmark appears under CoA 1, but we do not add plus signs under CoAs 2 and 3. This does not preclude application of these initiatives to other parts of the world.

Estimated resources for each CoA appear at the bottom of the matrix. For CoA 3, resources are divided between the first two (MC-12 and OA-X) and second two initiatives (gunship and mobility platform, which are bracketed). The total incremental manpower, total aircraft inventory, investment costs, and annual O&S costs refer to fully implemented initiatives. Appendix B details the breakout and flow of these manpower and cost figures.

CHAPTER FOUR

Recommendations for Successful Implementation of Irregular Warfare Initiatives

Four Requirements for Successful Organizational Change

Implementing the kinds of institutional, operational, and force structure changes we are proposing in our CoAs and initiatives will prove challenging. For the Air Force leadership to implement these changes, they would need to consider strategies for managing and improving large government and commercial organizations, as described in the management literature. According to this literature, successful change in a large organization such as the U.S. Air Force requires four factors:

- widespread belief in the need for change
- clear, sustained leadership, including support from top executives
- broad participation in diagnosing problems and planning the change
- flexible, incremental implementation, involving experimentation, feedback, adaptation, and building on prior success to institutionalize change.[1]

Our IW proposals take these factors into account.

The "widespread belief in the need for change" seems to be bubbling up from USAF tactical experiences but has been slow to spread to higher echelons in the Air Force. Over the past several years, and

[1] See Hal G. Rainey, *Understanding & Managing Public Organizations*, San Francisco: Jossey-Bass Publishers, 1997, p. 339.

certainly since the insurgency in Iraq gained steam, we have observed broad-based bottom-up support for change. But until recently, the top-down support that constitutes the second factor has been relatively weak. For this reason, we created the CoA 0 leadership initiatives that should help provide the basis for "clear, sustained leadership, including support from top executives."

The Air Force's IW Task Force and the Office of Air Force Lessons Learned activities are steps in the right direction since they encourage "broad participation in diagnosing problems and planning the change." Likewise, our consultations with a wide array of experts across the Air Force were aimed at encouraging broad participation in our study. Unfortunately, "planning the change" is complicated in the USAF due to the diversity of interests and communities ("tribes" in the vernacular) and the fiscal constraints that all but guarantee that one community's gain will be another's loss. Again, this is the role for leadership—they must establish and enforce priorities.

Finally, the need to approach change as an iterative learning and adaptation process is behind the structure and "vectoring" of our incremental CoAs. Choosing and implementing CoAs along these lines will not be a one-time event, but rather a long series of efforts that will require sustained involvement by the leadership and the flexibility to adapt and adjust programs over time. This is true both because the commitments outlined here have long time scales, and also because we recognize that our adversaries are clever and responsive, often on time scales less lengthy than ours. Hence, to confront and defeat these evolving threats most effectively, the USAF would need to implement the transformative processes in these CoAs with an eye toward adapting them to environments, threats, and demands that will likely change over time.

The Importance of Air Force Technology Leadership

Although we have focused a great deal of attention in this report on macro IW issues such as leadership, culture, and human capital, we are cognizant of the fact that the USAF is a leader and innovator in applying high technology to challenging, complex problems. We also understand that emerging operational concepts that utilize such technology have much to offer in the long war.

For example, certain information and sensor technologies, especially when mated with aircraft to provide access and persistent presence, can enable operations such as border monitoring and control, population monitoring, and general situational awareness that are critical to many IW operations. USAF cyber warriors can help inform and influence indigenous populations through directed psychological operations, while at the same time limiting an adversary's freedom of action in cyberspace through signals intelligence and electronic warfare means enabled by airborne access and persistence. Finally, limiting collateral damage is absolutely critical in IW. The USAF can be at the forefront of developing timely precision strike concepts enabled by

- situational awareness from space and airborne surveillance and reconnaissance platforms
- precision navigation enabled by the global positioning system
- networked sensors and shooters
- low-collateral-damage munitions that neutralize only their intended targets.

Of course, some of the most effective low-collateral-damage weapons may be nonkinetic. For example, jamming adversary communications can help protect Coalition forces and indigenous people (e.g., from remotely detonated improvised explosive devices) and can even support troops in contact through on-call jamming (e.g., "electronic warfare CAS"). The USAF can build on its existing competencies in surveillance and intelligence, especially signals intelligence, to exploit evolving information processing and communications technology for counterinsurgency purposes.

Because building partner capacity and operating indirectly are so central to irregular warfare, the Air Force could undertake a dedicated effort to make the fruits of these technologically driven concepts available to partners within export control parameters and without compromising sensitive or classified U.S. capabilities. In fact, concept development activities could explicitly consider applicability to capacity-building and foreign internal defense.

Many emerging concepts exploit cooperation with ground forces and/or administrative measures taken by the host nation. The Air Force generally cannot perform these missions on its own, and hence many promising areas of airpower application require cooperation with other services, U.S. government agencies, and partner governments. New applications may not fit well with any particular established role or mission. We therefore suggest joint and interagency investigation of innovative counterinsurgency technologies and operational concepts. Organizations and mechanisms to promote joint thinking with other services and other government agencies could be developed further.

It would be the responsibility of the irregular warfare integrator at Headquarters Air Force and the expanded CIWC to ensure that technological opportunities for novel IW applications are explored, developed, and fielded. In fact, over time, it might be appropriate for the USAF to create a significantly larger organization than the CIWC with a broader charter—perhaps an IW think tank.

CHAPTER FIVE

Conclusions

In summary, the DoD leadership has directed the Air Force and the other services to adapt to the emerging threat environment characterized as the "long war." The direction is clear but the specifics present numerous challenges, not the least of which is the need to begin to change the USAF culture from one focused on the challenges of major combat operations (which are not going away) to one equally adept in irregular warfare. The U.S. Air Force has a great deal to offer in this realm—indeed, U.S. efforts to counter terrorism and insurgency, support partners in foreign internal defense, and bring stability to vulnerable populations *could not be done without the Air Force.* It is in this spirit that we formulate and propose the courses of action in this report.

There is a compelling rationale for immediately beginning to implement the initiatives in CoAs 0, 1, and 2—or some versions thereof. The need to adapt as an institution, to redouble efforts in Iraq and Afghanistan, to pressure adversaries, and to support partners around the world is present, imperative, and enduring. We estimate that these first three CoAs would require about 7,000 personnel and $10 billion in investment and O&S funding out to 2015. Of course, there is still work to be done to refine, assess, resource, and execute the details of some initiatives, but the Air Force can build on an expanding body of background work and, especially, the operational and institutional experience among its members to pursue the vectors we have proposed here.

The initiatives in these CoAs offer an appropriate and strategically coherent mix of capabilities and capacities that would position the Air Force to meet the guidance set forth by DoD leadership. The foundational CoA 0 initiatives would ensure the development and nurturing of IW expertise and focused IW concepts, and would establish an enduring IW mindset in the Air Force from the leadership on down. By pursuing CoA 1 initiatives to push essential air capabilities forward to Iraq and Afghanistan, prepare for sustained operations and expanded Air Force roles there, and broaden efforts to build effective indigenous airpower, the Air Force would provide needed enhanced support to U.S. interests in those two countries while signaling its sense of urgency and responsiveness. Finally, adopting the initiatives in CoA 2 would set the Air Force on a course to meet demand in the rest of the world and to maintain pressure on terrorist and insurgent groups wherever they threaten U.S. interests.

As the Air Force implements CoAs 0, 1, and 2 and gains additional experience in and knowledge about IW operations, USAF leaders can take more time to consider the more ambitious and costly initiatives in CoA 3. It should be noted, however, that while the gunship and special operations forces mobility concepts in CoA 3 are by far the most costly initiatives in the study in terms of investment, the operational needs that they would meet—responsive, persistent, tailorable fires and clandestine infiltration/exfiltration in denied areas, respectively—present challenges that are no less pressing than the imperatives driving other CoAs.

Our rough cost estimates seem rather modest when compared with the USAF's top line budget and end strength, but in the emerging fiscally constrained environment, making even relatively modest changes in levels of resources can be challenging. As of fall 2008, supplemental funding acts for the Global War on Terrorism had provided the resources for many of the Air Force's IW-related activities, particularly in relation to training and operations in Iraq and Afghanistan. Supplemental funding will not be available indefinitely, and without dedicated funding sources, long-term planning for IW activities in the USAF will be impossible. Therefore, the USAF will have to perform trade studies to find the funding within its existing resources. If it is to move beyond ad hoc activities and embrace irregular warfare in its institutions, force planning, and operations, the Air Force must itself put up the resources. Money talks.

Finally, although there are material initiatives in these CoAs, many of the initiatives are nonmaterial. People, more than platforms, are the key to transforming the Air Force and rebalancing its IW and conventional capabilities, allowing the USAF to work more effectively and efficiently by, with, and through partner nations in the long war. The success of these CoAs and the specific initiatives will depend on inculcating an IW mindset across the Air Force, improving education and training, and demonstrating a commitment to promote and sustain the most qualified airmen for IW-related operations. As Secretary of Defense Gates wrote recently, "In the end, the military capabilities needed cannot be separated from the cultural traits and the reward structure of the institutions the United States has: the signals sent by what gets funded, who gets promoted, what is taught in the academies and staff colleges, and how personnel are trained."[1]

---

1 Gates, January/February 2009.

APPENDIX A

Snapshot Descriptions of the Irregular Warfare Initiatives

In this appendix we provide "snapshots" that describe each of the irregular warfare initiatives introduced in this monograph. On each page we state the objectives for the proposed initiative. Objectives stem from earlier work, which demonstrated to the satisfaction of both the Air Force and RAND that there are indeed pressing problems to which the objectives are appropriate. Numerous possibilities have been discussed and debated for each of the objectives—sometimes formally and sometimes informally. As befits the nature of this monograph, however, what follows moves directly to brief descriptions of the study's recommendations to meet the objectives.

Our recommendations are expressed primarily in terms that capture the essence of what must be accomplished, but they are also accompanied by first-cut estimates of resource requirements and specific measures. Such details may very well need to be reexamined in more depth, but the directions set are correct. Following the statement of objectives, we provide a general description of the proposed initiative, an explanation of employment and fielding concepts, and the estimated resources required along with a timeline. This information is of varying levels of fidelity and should be regarded as a first-order estimate. Subject matter experts in the Air Force should review each proposed initiative before decisions are made to move forward. Each snapshot identifies the CoA(s) with which the initiative is associated.

For each initiative we include a listing of manpower, total aircraft inventory, investment costs, annual costs for operations and support, and the years of initial capability (when manpower and/or platforms begin to come on line) and full capability (when final numbers of personnel and platforms are deployed).

We also give sources for the initiative. These are organizations that provided ideas and information and, in many cases, advocacy for the initiative (though not necessarily the details on resources and employment). In some cases there are multiple sources. Sometimes the source is identified as "ACC IW Workshop"—this refers to a two-day conference sponsored by Air Combat Command (ACC) and held at Langley Air Force Base on October 21–22, 2008, in which multiple major command, Air Staff, Secretariat, and other organizations participated.

Provide Leadership Emphasis

Objectives

- Emphasize the high priority that IW will have over the long term to deepen and expand IW capabilities in the Air Force
- Overcome the internal and external perception that IW is and will remain a low priority for the USAF
- Set the stage for IW as a priority in force planning, career management, assignments, training, and operations

Proposed Approach

- Convince institutional Air Force in speeches, policy guidance, and programmatic decisions that IW will remain a core USAF mission, even after a drawdown in Iraq
- CORONAs, communications to Air Force leadership, the Air Force as a whole, interactions with DoD
- Follow up with significant concrete steps (careers, organizations, funding)

Estimated Timeline and Costs

- CoA 0

Personnel	0
Total aircraft inventory	0
Investment	0
Annual O&S	0
Year of capability	
Initial	2009
Full	2009

Specific Proposals for Execution

- CoA 0 envisions the Secretary of the Air Force, Chief of Staff of the Air Force, and other USAF leaders emphasizing IW early and often

Sources

- Multiple USAF staffs; RAND

Designate Headquarters Air Force Integrator for Irregular Warfare

Objectives

- Address the lack of institutionalized leadership for USAF IW issues; many IW arrangements currently ad hoc and somewhat personality dependent
- Ensure development and nourishment of an IW mindset within the USAF

Proposed Approach

- Establish a single Air Staff integrator for USAF-wide IW activities
- Oversee IW strategy, doctrine, and resources; studies and analyses; IW training and designed operational capability statements; concept development; joint/interagency interaction (especially with the Army)
- Advocate for IW resources within the Air Force corporate structure (board, council, panel levels)

Estimated Timeline and Costs

- CoA 0

Personnel	50
Total aircraft inventory	0
Investment	$2 million
Annual O&S	$10 million
Year of capability	
Initial	2009
Full	2010

Specific Proposals for Execution

- CoA 0 has three options, each headed by a general officer:
  1. A new directorate in one of the Headquarters Air Force A-Staffs (e.g., A3/5)
  2. A new deputy chief of staff organization for IW (A-Staff "A11") headed by a two- or three-star general officer

3. A new division-plus sized organization headed by a two-star reporting directly to the Vice Chief of Staff of the Air Force

Sources

- RAND; ACC IW Workshop

Emphasize Irregular Warfare Career Incentives

Objectives

- Develop and nurture a long-term IW experience base in the USAF—especially for aviation advisors
- Reverse the perception that IW experience is a "kiss of death" for a USAF career

Proposed Approach

- Provide incentives and rewards for personnel to pursue IW experience in their Air Force careers
- Ensure that the right personnel are selected for advisory positions
- Track these personnel to ensure that the USAF can provide a flexible and "scalable" IW capability

Estimated Timeline and Costs

- CoA 0

Personnel	0
Total aircraft inventory	0
Investment	0
Annual O&S	0
Year of capability	
Initial	2009
Full	2010

Specific Proposals for Execution

- CoA 0 pursues multiple emphases:
  - Letters to promotion boards
  - Selection criteria for advisors
  - Special experience identifier tracking (in progress)
  - IW-related career requirements
  - High priority for assignments, including IW-related joint/interagency billets
  - Air National Guard involvement

Sources

- Multiple USAF staffs; Office of Air Force Lessons Learned; RAND

Expand Irregular Warfare Curricula in Education and Training

Objectives

- Mandate or increase exposure of airmen to IW concepts throughout their careers
- Build cadre of airmen who can be IW commanders and planners

Proposed Approach

- Integrate IW education into curricula beginning with accession, Reserve Officer Training Corps, Officer Training School, U.S. Air Force Academy, enlisted training, and professional military education
- Create IW campaign planning course to help prepare USAF planners to work with joint/interagency in theater and bring airpower most effectively and efficiently to bear in IW fights

Estimated Timeline and Costs

- CoA 0

Personnel	50
Total aircraft inventory	0
Investment	0
Annual O&S	$8 million
Year of capability	
Initial	2009
Full	2010

Specific Proposals for Execution

- CoA 0 adds IW-dedicated faculty to existing education and training venues; creates a one-week course on IW campaign planning (e.g., at USAF Special Operations School or Air University)

Sources

- ACC IW Workshop; multiple USAF staffs; Office of Air Force Lessons Learned

Expand Coalition Irregular Warfare Center

Objectives

- Provide a USAF focal point for IW-related concept development, technology exploitation
- Evaluate potential operational uses of emerging technologies
- Develop IW-specific tactics, techniques, and procedures
- Coordinate with RDT&E organizations (e.g., Defense Advanced Research Projects Agency, Air Force Research Laboratory, National Security Agency, Department of Energy laboratories, etc.)
- Work U.S./partner nation seams (interoperability, synergy, transferability, etc.)

Proposed Approach

- Build up current CIWC within the ACC/Air Warfare Center or create a new IW center
- Create permanent detachments at key major commands and other organizations (with personnel prepared to observe ongoing operations)
  - AFSOC, ACC, Air Mobility Command, Air Force Space Command, Air Force Materiel Command, AETC, Air Force Research Laboratory
  - Intelligence Community (Defense Intelligence Agency, Central Intelligence Agency, National Security Agency, etc.)

Estimated Timeline and Costs

- CoA 0

Personnel	81
Total aircraft inventory	0
Investment	$4 million
Annual O&S	$13 million
Year of capability	
Initial	2009
Full	2011

Specific Proposals for Execution

- CoA 0 adds 41 personnel to the 17 assigned (total of 58) at the CIWC and adds detachments of one to four airmen in each of eight to ten organizations

Sources

- ACC Air Warfare Center/CIWC; RAND

Embed Air Expertise Forward

Objectives

- Improve the effectiveness and efficiency of air support to distributed, decentralized ground operations in Iraq, Afghanistan
- Increase the focus on operational level of IW
- Adapt major combat operations processes to unique IW requirements
- Integrate CAS, lift, ISR, medical evacuation, information operations
- Designate senior IW-experienced airmen to work with equivalent U.S. Army (eventually Iraqi and Afghan) planners

Proposed Approach

- Assign senior officers and staffs in Baghdad and Kabul with planning and tasking authority, reachback to the CAOC
- Create IW "air effects" (AE) cells at corps/division and joint special operations task force levels, linked with air operations centers, to plan operations and coordinate lower-level planning
- Augment tactical air control parties in brigade combat teams with "air planning" (AP) cells to plan and synchronize all airpower contributions to ground operations

Estimated Timeline and Costs

- CoA 1

Personnel	262
Total aircraft inventory	0
Investment	$8 million
Annual O&S	$36 million
Year of capability	
Initial	2009
Full	2010

Specific Proposals for Execution

- CoA 1 envisions
  - Air commander/staffs at Multi-National Force–Iraq, Bagram Air Base (31)

- In Iraq: colonel-level AE director/staff at each of five divisions and joint special operations task force (42); lieutenant colonel–level AP director/staff at each of fifteen brigade combat teams in Iraq (105)
- In Afghanistan: colonel-level AE director/staff at each of three divisions and Joint Special Operations Task Force (21); lieutenant colonel–level AP director and staff at each of nine brigade combat teams in Afghanistan (63)

Sources

- 609th Air Operations Center; Office of Air Force Lessons Learned; RAND

Add Joint Terminal Attack Controllers

Objectives

- Ensure USAF ability to provide CAS and armed overwatch to U.S. and Iraqi ground forces during and following drawdown of U.S. brigade combat teams
- Reduce deployment stress on JTACs and related Air Force career fields
- Over time, allow brigade combat teams returning to home station to "bring their JTACs with them"

Proposed Approach

- With adequate ISR, communications, and U.S. joint fires observers at Iraqi battalion level, tactical air control parties can control CAS throughout Iraq
- Cover every U.S. mobile integrated tactical terminal at every Iraqi army division headquarters (13–16) plus Iraq air support operations center
- Training pipeline (to include live CAS sorties) will need to be expanded
- New counterinsurgency aircraft may help

Estimated Timeline and Costs

- CoA 1

Personnel	216
Total aircraft inventory	0
Investment	$5 million
Annual O&S	$22 million
Year of capability	
Initial	2009
Full	2010

Specific Proposals for Execution

- CoA 1 would field an additional 40 JTACs; 32 enlisted tactical air controllers (1C4)
- 1 air liaison officer, 2 JTACs, 2 enlisted tactical air controllers per division
- Provide 2:1 rotation base

Sources

- RAND

Push and Sustain Airborne ISR and PED

Objectives

- Improve the effectiveness and efficiency in Air Force ISR support to distributed ground operations in Iraq and Afghanistan
- Provide flexible, timely full-motion video and signals intelligence capabilities to ground forces
- Additional 10 orbits full time (24/7)
- Augment unmanned systems
- Alleviate the 1:1 dwell in USCENTCOM and build a cadre of MC-12 experts

Proposed Approach

- Office of the Secretary of Defense has given the USAF 37 MC-12 through Project Liberty (7 in the continental United States; 30 deployed)
- Force is not sustainable due to lack of counterterrorism aircraft for training
- Develop sustainable force of MC-12s, PED for USCENTCOM, other COCOMs
- Call for developing options for increasing rates of Predator MQ-1/9 deployment

Estimated Timeline and Costs

- CoA 1

Personnel	450
Total aircraft inventory	30
Investment	$460 M
Annual O&S	$93 M
Year of capability	
Initial	2010
Full	2012

CoA 3

Personnel	1,395
Total aircraft inventory	93
Investment	$1.4 B
Annual O&S	$289 M
Year of capability	
Initial	2012
Full	2016

Specific Proposals for Execution

- CoA 1: Add 30 aircraft (to current 37) to build to four squadrons with 15 primary assigned aircraft each

- CoA 3: Buy additional 93 MC-12s (for a total CoA 1 + 3 of 160, with 50 deployed, 100 for continuation training, 10 for testing), + 72 enlisted and officer intelligence specialists for PED

Sources

- Air Combat Command/A3F

Emphasize Innovative Information Operations

Objectives

- Better exploit key USAF technical capabilities in computer-network and electronic warfare to support the joint fight against terrorist and insurgent networks
- Counter adversary use of the Internet
- Exploit technology for directed psychological operations
- Develop "electronic warfare CAS" to provide timely electronic warfare support to ground forces (e.g., for local communications jamming and counter–improvised explosive device)

Proposed Approach

- Put dedicated information and electronic warfare airmen in the CAOC
- Fill USAF interagency billets (especially National Security Agency)

Estimated Timeline and Costs

- CoA 1

Personnel	54
Total aircraft inventory	0
Investment	$2 million
Annual O&S	$7 million
Year of capability	
Initial	2009
Full	2010

Specific Proposals for Execution

- CoA 1 envisions an information warfare flight of 24 people in the CAOC to support Iraq and Afghanistan, plus 30 USAF billets at the National Security Agency

Sources

- RAND; Office of Air Force Lessons Learned

Synchronize Strategic Communications

Objectives

- Assure that Air Force operations are consistent with national and COCOM strategic communications messages
- Advise Iraqi and Afghan people of the nature and purpose of air operations
  - Support/legitimize their governments
  - Defend them (and their security forces)
  - Help them (infrastructure, humanitarian operations)
  - Counter adversary propaganda
- Integrate strategic communications with operations

Proposed Approach

- Change CAOC strategic communications focus from targeting the U.S. public (by telling the Air Force "good news" story to people at home) to targeting the indigenous populations and legitimizing their governments
- Involve strategic communications expertise in operational planning

Estimated Timeline and Costs

- CoA 1

Personnel	0
Total aircraft inventory	0
Investment	0
Annual O&S	0
Year of capability	
Initial	2009
Full	2009

Specific Proposals for Execution

- CoA 1 envisions a refocusing and extension of the current strategic communications efforts in the CAOC

Sources

- Office of Air Force Lessons Learned; RAND

Mitigate Agile Combat Support Shortfalls

Objectives

- Ensure long-term ability to execute agile combat support tasking by security forces, maintenance, civil engineering, medical, and transportation support in Iraq and Afghanistan
- Rationalize presentation of ACS capabilities in support of building partner capacity, including base support, improved governance/services, and institution-building

Proposed Approach

- Reduce stress on key career fields from USAF and joint expeditionary, or in-lieu-of, tasking in Iraq and Afghanistan
- Assume full Reserve Component mobilization, 2:1 dwell
- Assume two options for security forces and maintenance ("manpower" or "nonmanpower")
- Develop ACS employment concepts for aviation advisory taskings elsewhere

Estimated Timeline and Costs

CoA 1		CoA 2	
Personnel	950	Personnel	0
Total aircraft inventory	0	Total aircraft inventory	0
Investment	$11 million	Investment	0
Annual O&S	$68 million	Annual O&S	0
Year of capability		Year of capability	
Initial	2009	Initial	2009
Full	2013	Full	2010

Specific Proposals for Execution

- CoA 1 increases manning and/or takes nonmanpower actions in stressed ACS career fields (assumes the least costly of two options; see footnote 8, p. 24)
- CoA 2 envisions concepts for ACS mobile training teams focused on institution-building, humanitarian aid, and medical and construction influence operations

Sources

- RAND; AF/A5RQ

Operationalize Analyses

Objectives

- Ensure that lessons *observed* in Iraq, Afghanistan become lessons *learned*
- Ensure tactical-level experience and analyses (such as those the Office of Air Force Lessons Learned has collected) are implemented in force planning and TTPs
- Institutionalize analysis processes based on top-down strategy guidance, direction, and resources

Proposed Approach

- Develop an analytic framework, processes, and manning to better exploit operational lessons
- Focus on operational-level measures of effectiveness—not just tactical-level measures of performance
- Improve data collection (on mission planning processes and inputs, tasking, execution, and assessment), especially data for air-ground measures of effectiveness
- Provide means to act on results (e.g., to adapt relevant processes)

Estimated Timeline and Costs

- CoA 1

Personnel	12
Total aircraft inventory	0
Investment	0
Annual O&S	$2 million
Year of capability	
Initial	2009
Full	2010

Specific Proposals for Execution

- CoA 1 would stand up a dedicated air-ground analysis cell in the CAOC with reachback support from Army and Air Force analysis cells in the continental United States

Sources

- RAND; Office of Air Force Lessons Learned

Expand Advisor Training

Objectives

- Ensure that advisory training for general-purpose forces is funded and institutionalized
- Include robust cultural/language skills training beyond familiarization
- Ensure that advisors are identified and tracked postdeployment, develop a career path
- Address command and control issues with the course and who funds equipment

Proposed Approach

- Call for GPF training funding in next Air Force Program Objective Memorandum
- Ensure a dedicated, funded aviation advisory training center of excellence
- Address issues with current training (i.e., command and control issues, number of training days, ensure new advisors are properly equipped)
- Systematically collect lessons from the field to improve training course
- Establish training standards for all advisors (including AFSOC candidates, who receive additional training as combat aviation advisors)

Estimated Timeline and Costs

- CoA 1

Personnel	20
Total aircraft inventory	20
Investment	$100 M
Annual O&S	$15 M
Year of capability	
Initial	2009
Full	2010

CoA 2

Personnel	25
Total aircraft inventory	30
Investment	$150 M
Annual O&S	$18 M
Year of capability	
Initial	2011
Full	2012

Specific Proposals for Execution

- CoA 1 envisions tailorable training pipeline to support 800 GPF advisors for Iraq and Afghanistan plus 400 potential AFSOC candidates
- CoA 2 expands pipeline to support additional 1,500 advisors in GPF advisory wing and air commando wing for steady state
- Adds instructors; assumes an advisors' career average of six years

Sources

- ACC IW Workshop; RAND

Stand Up General-Purpose Forces Advisor Unit and Elements

Objectives

- Enable the Air Force to support a persistent advisory presence in partner nations
- Ensure that partner nation demands are appropriately defined and implemented
- Provide a dedicated rotation base, continuation training, lessons learned, and career opportunities for permanent GPF air advisors

Proposed Approach

- Establish an operational organization for a long-term GPF advisory capability
- Dedicate airmen with advisory and regional expertise to capturing air-related demands and needs of partner nations in theater engagement plans and in tasking to USAF advisory units
- Scalable, reachback into rest of USAF
- Support USSOCOM security force assistance concept

Estimated Timeline and Costs

- CoA 2

Personnel	1,150
Total aircraft inventory	80
Investment	$1.3 billion
Annual O&S	$104 million
Year of capability	
Initial	2011
Full	2015

Specific Proposals for Execution

- CoA 2 envisions 30 theater air advisors in each of five COCOM headquarters; a Total Force GPF wing with COCOM-aligned squadrons and regionally oriented advisors from multiple career fields, with partner nation equipment and civil-military air infrastructure development kits (air traffic control radars, lights, etc.)
- Coordinated with National Guard State Partnership Program

Sources

- AF/A5R; AF/A5XS

Add Combat Aviation Advisors

Objectives

- Reduce mismatch between assessed demand for combat aviation advisors (CAA) and supply
- Synchronize AFSOC CAA capability and capacity with new role of USSOCOM as lead proponent of security forces assistance
- Help meet increasing demand for CAA-like skills in Iraq and Afghanistan

Proposed Approach

- Broaden combat aviation advisory capacity
- New USSOCOM security forces assistance concept calls for an additional 6th Special Operations Squadron–like CAA squadron

Estimated Timeline and Costs

- CoA 1

Personnel	230
Total aircraft inventory	20
Investment	$109 M
Annual O&S	$33 M
Year of capability	
Initial	2010
Full	2013

CoA 2

Personnel	500
Total aircraft inventory	40
Investment	$212 M
Annual O&S	$66 M
Year of capability	
Initial	2011
Full	2016

Specific Proposals for Execution

- CoA 1 creates a second AFSOC CAA squadron to support USSOCOM security forces assistance, advanced training in Iraq, Afghanistan
- CoA 2 creates air commando wing of five squadrons, each aligned to a COCOM, and includes Air National Guard units

Sources

- RAND; Air Force Special Operations Command

Expand Air Force Human Intelligence

Objectives

- Expand and better exploit the human intelligence potential in the Air Force
  - Country Teams (defense attachés and security assistance officers)
  - COCOM theater engagement plan teams
  - GPF and special operations advisors
  - State Partnership Program exchanges
- Augment foreign area officer cadre and fully staff Air Force billets at the Department of Homeland Security

Proposed Approach

- Provide education and training for all airmen in positions with human intelligence opportunities
- Create processes to exploit their efforts (human intelligence PED)

Estimated Timeline and Costs

- CoA 2

Personnel	20
Total aircraft inventory	0
Investment	0
Annual O&S	$3 million
Year of capability	
Initial	2010
Full	2011

Specific Proposals for Execution

- CoA 2 creates and vets "every airman is a (human intelligence) sensor" concepts; develops specialized education and training programs

Sources

- RAND; AFSOC

Build Regional Air Academies

Objectives

- Enhance and sustain Air Force investments in tactical training of foreign units
- Establish enduring relationships with future partner air arm leaders
- Meet substantial current and future demand for individual military education and training for Iraq, Afghanistan, and other priority partners

Proposed Approach

- Using the Inter-American Air Force Academy as a template, establish new international military education and training institutions for other COCOMs providing instrument and other flight training, maintenance, airbase defense, logistics courses
- Each with 50 faculty, 50 other staff with regional expertise, mobile training team capacity
- Each with 500 students, primarily from middle ranks

Estimated Timeline and Costs

CoA 1		CoA 2	
Personnel	100	Personnel	300
Total aircraft inventory	0	Total aircraft inventory	0
Investment	$14 M	Investment	$15 M
Annual O&S	$13 M	Annual O&S	$40 M
Year of capability		Year of capability	
Initial	2010	Initial	2011
Full	2011	Full	2013

Specific Proposals for Execution

- CoA 1 establishes institute for USCENTCOM emphasizing Iraqi and Afghan needs (in addition to existing SOUTHCOM academy)
- CoA 2 establishes institutes for U.S. Africa Command, U.S. Pacific Command, U.S. European Command

Sources

- ACC IW Workshop; AFSOC

Provide Transferable, Counterinsurgency-Dedicated Close Air Support and Armed Overwatch Platform

Objectives

- Provide a light attack and ISR capability in support of U.S. and partner ground forces that is transferable to partner nations
- Help enable persistent presence and engagement with partners

Proposed Approach

- USAF must first undertake an analytic assessment of the need for, and nature of, an Air Force–flown, counterinsurgency-dedicated CAS platform
- For illustrative purposes, this initiative envisions an OA-X
- Flexible, easy for partner air arms to operate

Estimated Timeline and Costs

- CoA 1

Personnel	1,000
Total aircraft inventory	100
Investment	$1.1 B
Annual O&S	$103 M
Year of capability	
Initial	2010
Full	2015

CoA 3

Personnel	3,000
Total aircraft inventory	300
Investment	$3.2 B
Annual O&S	$309 M
Year of capability	
Initial	2012
Full	2021

Specific Proposals for Execution

- CoA 1 would deploy about 100 aircraft (one replacement training unit and six operational squadrons to support two deployed locations); support GPF/CAA advisory and JTAC training
- CoA 3 would, depending on assessed need, deploy additional 300 aircraft (total of 400: two replacement training units, 12 operational squadrons) in multiple areas of operations

Sources

- ACC/A3F

Develop and Procure Light Cargo Aircraft

Objectives

- Help partners extend governance to ungoverned regions, provide aid (e.g., support country teams)
- Build partner capacity with locally unobtrusive, transferable platform
- Support distributed U.S. and partner military operations, including resupply and medical evacuation
- Right-size to other services' expansion of GPF partner-building capacity

Proposed Approach

- The light cargo aircraft would be a family of utility and cargo aircraft (a Small Multi-Mission Air Fleet) in the 3,000- to 6,000-pound payload range suited to specific regions
- It would be modular to enable expansion to medical evacuation, ISR capabilities
- May be patterned after AFSOC's Non-Standard Aviation program

Estimated Timeline and Costs

CoA 1		CoA 2	
Personnel	300	Personnel	1,000
Total aircraft inventory	30	Total aircraft inventory	105
Investment	$181 M	Investment	$635 M
Annual O&S	$41 M	Annual O&S	$142 M
Year of capability		Year of capability	
Initial	2010	Initial	2011
Full	2012	Full	2015

Specific Proposals for Execution

- CoA 1 would provide 30 aircraft (buy or lease) in two squadrons for use in Iraq and Afghanistan (with appropriate service ceilings for Afghanistan)
- CoA 2 would add 105 aircraft (two replacement training units in the continental United States, one operational squadron in each of the five COCOM areas of operation)

- Flyaway cost per light cargo aircraft is $6 million

Sources

- ACC IW Workshop; RAND

Develop and Deploy the Next-Generation Gunship

Objectives

- Enable gunship support (including responsive, tailorable, precise, sustained fires/operations) day and night in future semipermissive (i.e., man-portable air defense system and anti-aircraft artillery) environments

Proposed Approach

- Options include manned or unmanned, current, or next-generation platforms
- May leverage next-generation bomber to enable provision of gunship capabilities
- Includes standoff precision-guided munitions and ISR concepts
- Analysis of alternatives completed in 2005 (RAND)

Estimated Timeline and Costs

- CoA 3

Personnel	720
Total aircraft inventory	24
Investment	$11.2 billion
Annual O&S	$113 million
Year of capability	
Initial	2020
Full	2025

Specific Proposals for Execution

- To be determined by the Air Force, but for the purposes of this study we used next-generation gunship analysis of alternatives of 24 stealthy, manned gunships

Sources

- AFSOC; RAND

Develop and Deploy Next-Generation SOF Mobility Platform

Objectives

- Future capability to conduct infiltration/exfiltration in denied or politically sensitive areas over great distances
- Provide SOF rapid, self-deployable, global, high-threat, anti-access capability with agility in the objective area

Proposed Approach

- Baseline concept: a high-speed, long-range air mobility platform with vertical/short takeoff and landing capability and low-observable/stealth technology
- Analysis of alternatives completed in 2004 (AFSOC)

Estimated Timeline and Costs

- CoA 3

Personnel	600
Total aircraft inventory	24
Investment	$9.2 billion
Annual O&S	$103 million
Year of capability	
Initial	2025
Full	2029

Specific Proposals for Execution

- To be determined by the Air Force, but assumed to be a stealthy platform

Sources

- AFSOC

APPENDIX B

Estimates of Resources Required for Irregular Warfare Initiatives

This appendix depicts estimates of resources required for each of the proposed IW initiatives. Initiatives stem from earlier work, which demonstrated to the satisfaction of both the Air Force and RAND that there are indeed pressing problems appropriately addressed by the initiatives. Numerous possibilities have been discussed and debated for each of the initiatives—sometimes formally and sometimes informally. What follows moves directly to rough estimates for each initiative when fully implemented, and these represent the sum of the totals required for each applicable initiative. While some initiatives have logical associated offsets (e.g., next-generation gunship, SOF mobility platform, etc.), credit for these offsets is not taken here since adjudication of offsets for dollars and manpower should be within the context of trades among all the elements of the entire force. This is beyond the scope of our tasking. Flow of these estimates over time is provided in subsequent charts.

For each initiative, we calculate

- total aircraft inventory
- manpower
- RDT&E
- initial procurement
- military construction (MILCON)
- total investment cost (sum of RDT&E, initial procurement, and MILCON)
- annual operations and maintenance (O&M) costs

- annual manpower costs
- annual O&S costs (sum of O&M and manpower costs).

Costs are calculated in 2009 constant dollars.

We compute O&M costs for organizations by starting with the size of the manpower pool in the organization to be supported. Travel and office equipment costs, as well as supplies, are estimated based on the mission and size of the organization. If flying is involved, we apply O&M costs for the particular type or class of equipment using standard estimating techniques.

We calculate manpower costs using the total size of the manpower pool in the organization, an assumed distribution in rank/pay grade (unique for each initiative, based on mission), and DoD's tables for pay, allowances, and support by pay grade.

Investment costs include RDT&E, procurement, and MILCON estimates where applicable. RDT&E and procurement costs are estimated for the equipment associated with the initiative. In most cases, the equipment exists today; thus, we used "open market" costs and adjusted for specific configuration differences. In the case of next-generation gunship and SOF mobility initiatives, costs are estimated using information from analyses of alternatives on the next-generation gunship and the so-called "M-X" SOF mobility aircraft conducted in 2004. We calculate MILCON based on typical experience without benefit of site surveys.

Table B.1 details the estimated resources required for each initiative. The initiatives are listed in the order they appear in Table 3.1. However, initiatives that appear in two CoAs here appear in two rows. These initiative titles are followed by a number that denotes the CoA in which it appears. For example, the MC-12 initiative appears in CoAs 1 and 3; in Table B.1, "push and sustain airborne ISR and PED (1)" refers to the 30 MC-12s called for in CoA 1, and "push and sustain airborne ISR and PED (3)" refers to the additional 93 MC-12s called for in CoA 3.

Table B.1
Rough Initiative Estimates

Initiative	Total Aircraft Inventory	Man-power	RDT&E ($09M)	Procure-ment ($09M)	MILCON ($09M)	Total Investment ($09M)	Annual O&M ($09M)	Annual Pers Cost ($09M)	Total Annual ($09M)
Institutional									
Designate HAF integrator for IW	0	50	0	3	0	3	3	7	10
Emphasize IW career incentives	0	0	0	0	0	0	0	0	0
Expand IW curricula in education and training	0	50	0	0	0	0	1	7	8
Expand CIWC	0	81	0	0	4	4	1	12	13
Operational Capability									
Embed air expertise forward	0	262	0	8	0	8	3	34	36
Add JTACs	0	216	0	5	0	5	1	21	22
Push and sustain airborne ISR and PED (1)	30	450	0	456	4	460	65	28	93
Push and sustain airborne ISR and PED (3)	93	1,395	0	1,432	4	1,436	201	88	289
Emphasize innovative information operations	0	54	0	2	0	2	0	6	7
Synchronize strategic communications	0	0	0	0	0	0	0	0	0
Mitigate ACS shortfalls (1)	0	950	0	11	0	11	5	63	68
Mitigate ACS shortfalls (2)	0	0	0	0	0	0	0	0	0
Operationalize analyses	0	12	0	0	0	0	0	1	2

Table B.1 (continued)

Initiative	Total Aircraft Inventory	Man-power	RDT&E ($09M)	Procure-ment ($09M)	MILCON ($09M)	Total Investment ($09M)	Annual O&M ($09M)	Annual Pers Cost ($09M)	Total Annual ($09M)
Building Partner Capacity									
Expand advisor training (1)	20	20	0	100	0	100	12	3	15
Expand advisor training (2)	30	25	0	150	0	150	15	3	18
Stand up GPF advisor unit and elements	80	1,150	0	1,267	3	1,270	32	72	104
Add CAAs (1)	20	230	0	105	4	109	10	23	33
Add CAAs (2)	40	500	0	208	4	212	20	46	66
Expand USAF HUMINT	0	20	0	0	0	0	0	3	3
Build regional air academies (1)	0	100	0	1	13	14	1	13	13
Build regional air academies (2)	0	300	0	2	13	15	2	38	40
Future Procurement									
Provide transferable, COIN-dedicated CAS and armed overwatch platform (3)	300	3,000	0	3,228	0	3,228	120	189	309
Provide transferable, COIN-dedicated CAS and armed overwatch platform (1)	100	1,000	0	1,046	10	1,056	40	63	103
Develop and procure light cargo aircraft (1)	30	300	0	181	0	181	23	19	41
Develop and procure light cargo aircraft (2)	105	1,000	0	379	0	379	79	63	142
Develop and deploy the next-generation gunship	24	720	2,750	8,496	8	11,254	60	53	113
Develop and deploy next-generation SOF mobility platform	24	600	2,700	6,528	0	9,228	60	43	103
Totals	**896**	**12,485**	**5,450**	**23,610**	**67**	**29,127**	**751**	**899**	**1,650**

Figure B.1 presents the additional total aircraft inventory over time associated with the initiatives that call for procurement and deployment of aircraft. The time phasing is reasonably front-loaded and is based on judgment regarding how fast each initiative might be implemented when accorded a high priority within a constrained funding environment.

Figure B.2 depicts additional manpower over time. As with the total aircraft inventory, the time phasing is reasonably front-loaded

Figure B.1
Increase in Total Aircraft Inventory for All Initiatives

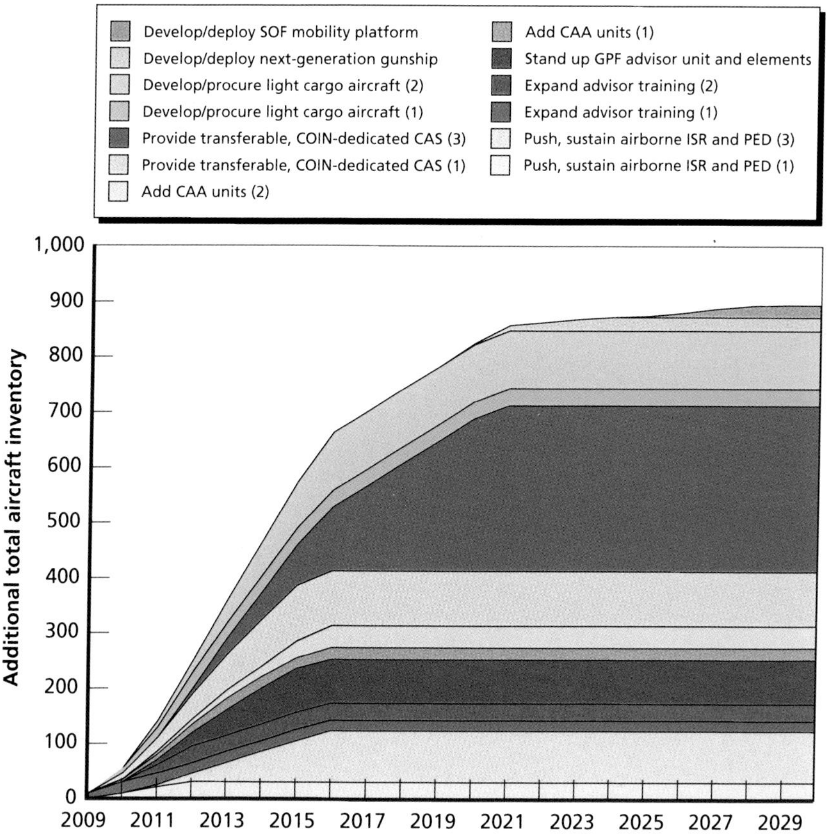

RAND *MG913-B.1*

and is based on judgment regarding how fast each initiative might be implemented when accorded a high priority within a constrained funding environment.

Figure B.2
Increase in Manpower for All Initiatives

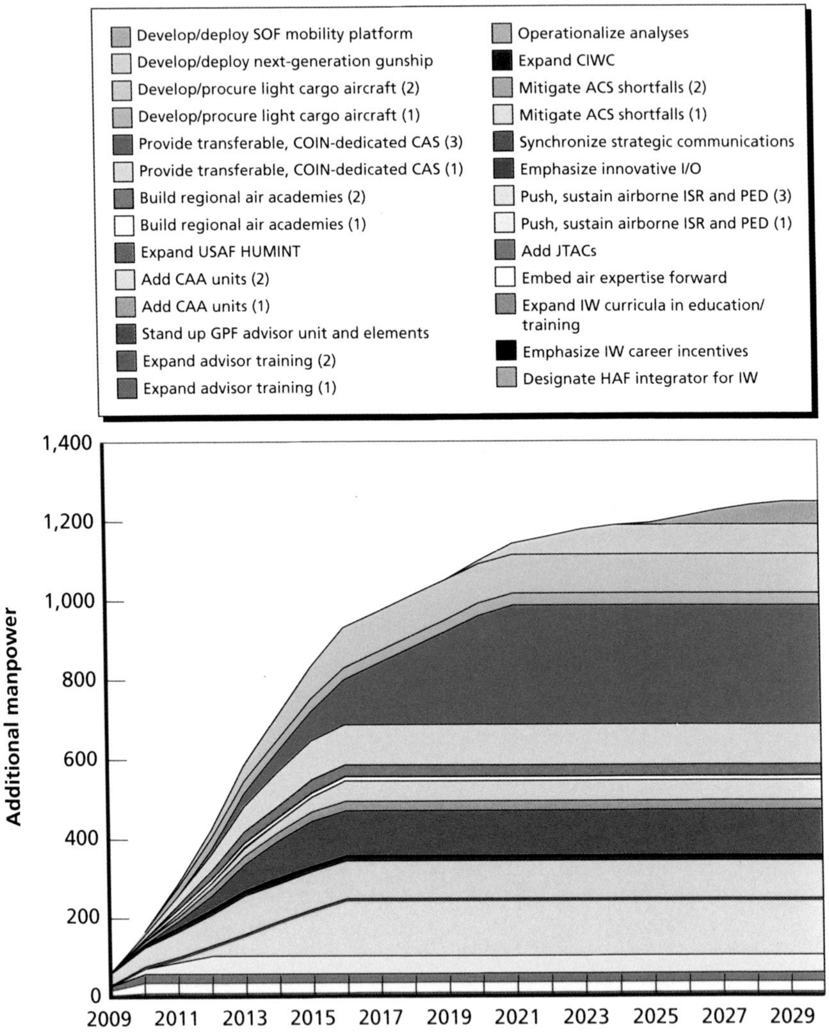

RAND *MG913-B.2*

Figure B.3 presents the added total costs over time with the same rules of thumb as total aircraft inventory and manpower. The exceptions are the two costliest initiatives, "Develop and deploy next-generation

Figure B.3
Estimated Costs for All Initiatives

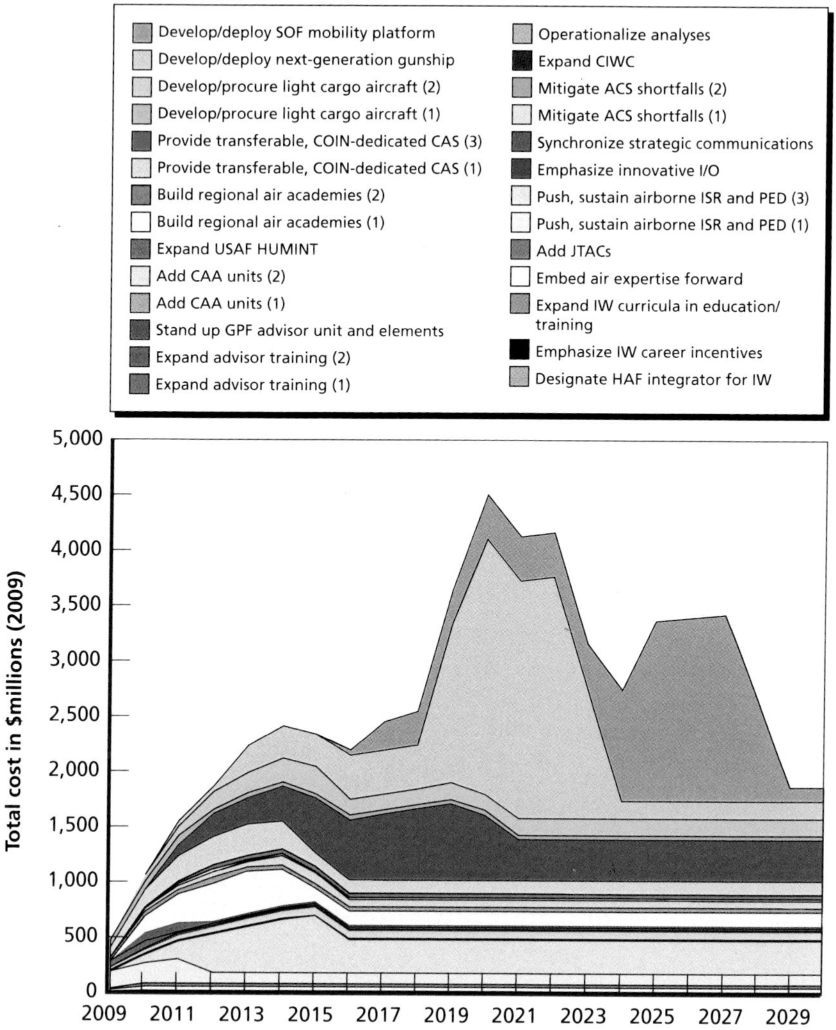

RAND *MG913-B.3*

gunship" and "Develop and deploy SOF mobility platform." We phased these two initiatives to relieve the stress on the budget around 2020, with the gunship preceding the mobility platform.

Bibliography

Following is a selective list of joint, U.S. Air Force, RAND, and other publications. This list is not intended to be exhaustive of publications related to irregular warfare or Iraq and Afghanistan.

Joint Publications

Defense Security Cooperation Agency, *Irregular Warfare (IW) Joint Operating Concept (JOC),* Version 1.0, Washington, D.C., September 11, 2007.

———, "Building Partner Capacity: Security Cooperation Tools," briefing, Washington, D.C., May 15, 2008.

———, *National Defense Strategy,* Washington, D.C., June 2008.

———, *Irregular Warfare (IW),* DoDD 3000.07, Washington, D.C.: Office of the Under Secretary of Defense for Policy, December 1, 2008.

Department of Defense, "DoD News Briefing with Lt. Gen. Odierno from Iraq," January 17, 2008. As of September 8, 2009: http://www.defenselink.mil/transcripts/transcript.aspx?transcriptid=4122

Gates, Robert M., "A Balanced Strategy: Reprogramming the Pentagon for a New Age," *Foreign Affairs,* January/February 2009. As of September 8, 2009: http://www.foreignaffairs.org/20090101faessay88103/robert-m-gates/how-to-reprogram-the-pentagon.html

———, transcript of speech at the National Defense University, September 29, 2008. As of September 8, 2009: http://www.defenselink.mil/speeches/speech.aspx?speechid=1279

Odierno, Raymond, Gen., "Multi-National Force–Iraq Commander's Counterinsurgency Guidance," Headquarters Multi-National Force–Iraq, Baghdad, Iraq, September 16, 2008.

Rumsfeld, Donald, U.S. Secretary of Defense, *Quadrennial Defense Review Report,* Washington, D.C.: Department of Defense, February 6, 2006.

United States Army, *Counterinsurgency*, Washington, D.C.: Headquarters, Department of the Army, FM 3-24/MCWP 3-33.5, December 2006.

U.S. Air Force Publications

Air Combat Command, *Irregular Warfare Air Combat Command Operating Concept*, Langley Air Force Base, Va., July 25, 2008; not available to the general public.

Air Force Special Operations Command, *Advanced Special Operations Air Mobility Platform (M-X) Analysis of Alternatives, Volume I*, September 30, 2004; not available to the general public.

———, "Combat Aviation Advisors," briefing, 6th Special Operations Squadron, April 2006.

———, *USAF Irregular Warfare Concept*, White Paper, May 2007.

Air Mobility Command, *Air Mobility Command on Irregular Warfare*, White Paper, Version 1.5, Scott Air Force Base, Ill., undated.

Childress, Timothy, Maj, USAF, "Train, Advise, Assist," concept briefing, Washington, D.C.: AF/A5XS, May 19, 2008.

———, "USAF Theater Air Advisory Squadron Concept," Washington, D.C.: Headquarters, United States Air Force, Directorate of Operational Planning, Policy and Strategy, Concepts, Strategy, and Wargaming Division, June 29, 2008.

Crowder, Gary, Col, "Air-Ground Integration in the 21st Century," documented briefing, Al-Udeid Air Base, Qatar: USAFCENT CAOC (609th Air Operations Center), April 2008.

Hansen, Blair, Brig Gen, "Project Liberty RC-12," briefing, Washington, D.C.: Headquarters, United States Air Force, Directorate of Intelligence, Surveillance, and Reconnaissance Capabilities, August 17, 2008.

Mueller, Stephen, Maj Gen, "IW Plan of Action and Milestones," briefing, Washington, D.C.: Headquarters, United States Air Force, Directorate of Operational Capability Requirements, September 24, 2008.

———, "Force Structure for the Long War," briefing, Washington, D.C.: Headquarters, United States Air Force, Directorate of Operational Capability Requirements, November 5, 2008.

Office of Air Force Lessons Learned, *Integration of Airpower in Operational Level Planning Report*, Washington, D.C.: Headquarters, United States Air Force, Directorate of Studies and Analyses, Assessment and Lessons Learned, August 22, 2008.

———, *Airpower in Irregular Warfare*, Washington, D.C.: Headquarters, United States Air Force, Directorate of Studies and Analyses, Assessment and Lessons Learned, September 25, 2008.

Pietrucha, Michael, Lt Col, "ACC and Irregular Warfare," briefing, presented at 2008 CAF/MAF Commanders Conference, undated.

Torres-Laboy, J. David, Lt Col, *A New Light Attack Aircraft: Making the Case for the Current Fight and Preparing for Future Conflicts*, Langley Air Force Base, Va.: Joint Air Ground Combat Division, ACC White Paper, March 2008.

———, and Lt Col Michael Pietrucha, *Finding a Square Peg for a Square Hole—IW and the Combat Air Force*, Version 1.6, Langley Air Force Base, Va.: Joint Air Ground Combat Division, ACC White Paper, May 2008.

United States Air Force, *Air Force Global Partnership Strategy: Building Partnerships for the 21st Century*, draft, Version 7.1, undated.

———, *Irregular Warfare*, AFDD 2-3, Washington, D.C.: Headquarters, Air Force, August 1, 2007.

———, *The 21st Century Air Force: Irregular Warfare White Paper*, Version 3.1, Washington, D.C.: Headquarters, Air Force, April 7, 2008.

———, *United States Air Force Irregular Warfare Roadmap FY09–FY12*, Version 1.6, Washington, D.C.: Headquarters, United States Air Force, Directorate of Operational Capabilities and Requirements, July 2008.

———, *U.S. Air Force Irregular Warfare Operating Concept*, Washington, D.C.: Headquarters, United States Air Force, Directorate of Operational Planning, Policy and Strategy, Concepts, Strategy and Wargaming Division, December 1, 2008.

RAND Publications

Dobbins, James, John G. McGinn, Keith Crane, Seth G. Jones, Rollie Lal, Andrew Rathmell, Rachel M. Swanger, and Anga R. Timilsina, *America's Role in Nation-Building: From Germany to Iraq*, Santa Monica, Calif.: RAND Corporation, MR-1753-RC, 2003.
http://www.rand.org/pubs/monograph_reports/MR1753/

Dobbins, James, Seth G. Jones, Keith Crane, and Beth Cole DeGrasse, *The Beginner's Guide to Nation-Building*, Santa Monica, Calif.: RAND Corporation, MG-557-SRF, 2007.
http://www.rand.org/pubs/monographs/MG557/

Dunham-Scott, Diana, Richard Mesic, and Edward O'Connell, "Implications from Iraq for Air Force Irregular Warfare Employment: Theater Air Control System," unpublished RAND research, May 2008.

Gompert, David, John Gordon, IV, Adam Grissom, David R. Frelinger, Seth G. Jones, Martin C. Libicki, Edward O'Connell, Brooke Stearns Lawson, and Robert E. Hunter, *War by Other Means: Building Complete and Balanced Capabilities for Counterinsurgency: RAND Counterinsurgency Study—Final Report*, Santa Monica, Calif.: RAND Corporation, MG-595/2-OSD, 2008.
http://www.rand.org/pubs/monographs/MG595.2/

Grissom, Adam, and David Ochmanek, *Train, Equip, Advise, Assist: The USAF and the Indirect Approach to Countering Terrorist Groups Abroad,* Santa Monica, Calif.: RAND Corporation, MG-699-AF, May 2008; not available to the general public.

Hamilton, Thomas, Brian G. Chow, Paul Dreyer, Gordon Bitko, and John E. Peters, "New Approaches to USAF Counterinsurgency Operations," unpublished RAND research, November 2008.

Jacobs, Jody, Gary McLeod, and Eric V. Larson, *Enhancing the Integration of Special Operations and Conventional Air Operations: Focus on the Air-Surface Interface,* Santa Monica, Calif.: RAND Corporation, MG-535-AF, 2007; not available to the general public.

Jacobs, Jody, Leland Joe, David Vaughan, Diana Dunham-Scott, Lewis Jamison, and Michael Webber, *Technologies and Tactics for Improved Air-Ground Effectiveness,* Santa Monica, Calif.: RAND Corporation, MG-573-AF, 2008; not available to the general public.

Johnson, David E., *Learning Large Lessons: The Evolving Roles of Ground Power and Air Power in the Post–Cold War Era*, Santa Monica, Calif.: RAND Corporation, MG-405-1-AF, 2007.
http://www.rand.org/pubs/monographs/MG405.1/

Jones, Seth, *Counterinsurgency in Afghanistan: RAND Counterinsurgency Study, Volume 4,* Santa Monica, Calif.: RAND Corporation, MG-595-OSD, 2008.
http://www.rand.org/pubs/monographs/MG595/

Lambeth, Benjamin S., *Air Power Against Terror: America's Conduct of Operation Enduring Freedom,* Santa Monica, Calif.: RAND Corporation, MG-166-1-CENTAF, 2006.
http://www.rand.org/pubs/monographs/MG166-1/

Libicki, Martin C., David C. Gompert, David R. Frelinger, and Raymond Smith, *Byting Back: Regaining Information Superiority Against 21st-Century Insurgents,* Santa Monica, Calif.: RAND Corporation, MG-595/1-OSD, 2007.
http://www.rand.org/pubs/monographs/MG595.1/

Long, Austin, *On "Other War": Lessons Learned from Five Decades of RAND Counterinsurgency Research*, Santa Monica, Calif.: RAND Corporation, MG-482-OSD, 2006.
http://www.rand.org/pubs/monographs/MG482/

Mesic, Richard, Edward O'Connell, and Dale Stahl, "The Implications of Irregular Warfare for the U.S. Air Force," unpublished RAND research, July 2007.

———, Robert Anderson, Myron Hura, Philip Anton, and James Chiesa, *Project AIR FORCE Perspectives on Air Force Information Warfare: A Summary Report*, Santa Monica, Calif.: RAND Corporation, MG-129-AF, 2004; not available to the general public.

Moore, Richard M., John Stillion, Clarence R. Anderegg, Susan R Bohandy, Michael Boito, James L. Bonomo, Michael Brown, Katherine M. Calef, James S. Chow, Leo C. Cloutier, Toby Edison, Scot Eisenhard, David R. Frelinger, Mark Gabriele, Roy O. Gates, Elham Ghashghai, Jon Grossman, Jeff Hagen, Gail Halverson, Thomas J. Herbert, Walter Hobbs, Alexander C. Hou, Phyllis Kantar, Michael Kennedy, Beth Lachman, Robert S. Leonard, David M. Matonick, Daniel Norton, David T. Orletsky, Harold Scott Perdue, Kevin Pollpeter, Robert Preston, Randall Steeb, Carl W. Stephens, Fred Timson, Michael S. Tseng, and Barry A. Wilson, *Next-Generation Gunship Analysis of Alternatives: Final Report—Volume I*, Santa Monica, Calif.: RAND Corporation, MR-1795-AF, 2004; not available to the general public.

Ochmanek, David, *Military Operations Against Terrorist Groups Abroad: Implications for the United States Air Force*, Santa Monica, Calif.: RAND Corporation, MR-1738-AF, 2003.
http://www.rand.org/pubs/monograph_reports/MR1738/

Owen, Robert C., and Karl P. Mueller, *Airlift Capabilities for Future U.S. Counterinsurgency Operations*, Santa Monica, Calif.: RAND Corporation, MG-565-AF, 2007.
http://www.rand.org/pubs/monographs/MG565/

Pirnie, Bruce R., and Edward O'Connell, *Counterinsurgency in Iraq (2003–2006): RAND Counterinsurgency Study—Volume 2*, Santa Monica, Calif.: RAND Corporation, MG-595/3-OSD, 2008.
http://www.rand.org/pubs/monographs/MG595.3/

Pirnie, Bruce R., Alan J. Vick, Adam Grissom, Karl P. Mueller, and David T. Orletsky, *Beyond Close Air Support: Forging a New Air-Ground Partnership*, Santa Monica, Calif.: RAND Corporation, MG-301-AF, 2005.
http://www.rand.org/pubs/monographs/MG301/

Stanley, William, James S. Chow, Jeff Hagen, Daniel M. Norton, David Vaughan, John Tonkinson, Gregory G. Hildebrandt, and Joel Kvitky, "Aircraft Systems for the U.S. Air Force and for Friendly States Facing Insurgencies: Executive Summary," unpublished RAND research, February 2007.

Stillion, John, David E. Thaler, Alexander C. Hou, David E. Johnson, Richard Mesic, Karl P. Mueller, and John E. Peters, "Air Force Phase Zero Capabilities-Based Analyses: Functional Area Analysis and Functional Needs Analysis," unpublished RAND research, September 2007.

Thaler, David E., Theodore W. Karasik, Dalia Dassa Kaye, Jennifer D. P. Moroney, Frederic Wehrey, Obaid Younossi, Farhana Ali, and Robert A. Guffey, *Future U.S. Security Relationships with Iraq and Afghanistan: U.S. Air Force Roles*, Santa Monica, Calif.: RAND Corporation, MG-681-AF, 2008. http://www.rand.org/pubs/monographs/MG681/

Vick, Alan J., Adam Grissom, William Rosenau, Beth Grill, and Karl P. Mueller, *Air Power in the New Counterinsurgency Era: The Strategic Importance of USAF Advisory and Assistance Missions*, Santa Monica, Calif.: RAND Corporation, MG-509-AF, 2006.
http://www.rand.org/pubs/monographs/MG509/

Other Publications

"Agreement Between the United States of America and the Republic of Iraq on the Withdrawal of United States Forces from Iraq and the Organization of Their Activities During Their Temporary Presence in Iraq," November 17, 2008. As of September 8, 2009:
http://graphics8.nytimes.com/packages/pdf/world/20081119_SOFA_FINAL_AGREED_TEXT.pdf

"A Conversation with Iraq's Prime Minister Nuri al-Maliki," United States Institute of Peace, July 23, 2009. As of September 2009:
http://www.usip.org/programs/initiatives/conversation-iraq-s-prime-minister-nuri-al-maliki

Luman, Ronald, Executive Editor, *Proceedings on Combating the Unrestricted Warfare Threat: Integrating Strategy, Analysis, and Technology*, 2008 Unrestricted Warfare Symposium, March 10–11, 2008.

Rainey, Hal G., *Understanding & Managing Public Organizations*, San Francisco, Calif.: Jossey-Bass Publishers, 1997.